A. F. Rocha, E. Massad, A. Pereira Jr.
The Brain: From Fuzzy Arithmetic to Quantum Computing

Studies in Fuzziness and Soft Computing, Volume 165

Editor-in-chief
Prof. Janusz Kacprzyk
Systems Research Institute
Polish Academy of Sciences
ul. Newelska 6
01-447 Warsaw
Poland
E-mail: kacprzyk@ibspan.waw.pl

Further volumes of this series
can be found on our homepage:
springeronline.com

Vol. 149. J.J. Buckley
Fuzzy Statistics, 2004
ISBN 3-540-21084-9

Vol. 150. L. Bull (Ed.)
Applications of Learning Classifier Systems,
2004
ISBN 3-540-21109-8

Vol. 151. T. Kowalczyk, E. Pleszczyńska,
F. Ruland (Eds.)
*Grade Models and Methods for Data
Analysis*, 2004
ISBN 3-540-21120-9

Vol. 152. J. Rajapakse, L. Wang (Eds.)
*Neural Information Processing: Research
and Development*, 2004
ISBN 3-540-21123-3

Vol. 153. J. Fulcher, L.C. Jain (Eds.)
Applied Intelligent Systems, 2004
ISBN 3-540-21153-5

Vol. 154. B. Liu
Uncertainty Theory, 2004
ISBN 3-540-21333-3

Vol. 155. G. Resconi, J.L. Jain
Intelligent Agents, 2004
ISBN 3-540-22003-8

Vol. 156. R. Tadeusiewicz, M.R. Ogiela
Medical Image Understanding Technology,
2004
ISBN 3-540-21985-4

Vol. 157. R.A. Aliev, F. Fazlollahi, R.R. Aliev
*Soft Computing and its Applications in
Business and Economics*, 2004
ISBN 3-540-22138-7

Vol. 158. K.K. Dompere
*Cost-Benefit Analysis and the Theory
of Fuzzy Decisions – Identification and
Measurement Theory*, 2004
ISBN 3-540-22154-9

Vol. 159. E. Damiani, L.C. Jain, M. Madravia
Soft Computing in Software Engineering,
2004
ISBN 3-540-22030-5

Vol. 160. K.K. Dompere
*Cost-Benefit Analysis and the Theory
of Fuzzy Decisions – Fuzzy Value Theory*,
2004
ISBN 3-540-22161-1

Vol. 161. N. Nedjah, L. de Macedo Mourelle
(Eds.)
Evolvable Machines, 2005
ISBN 3-540-22905-1

Vol. 162. N. Ichalkaranje, R. Khosla, L.C.
Jain
Design of Intelligent Multi-Agent Systems,
2005
ISBN 3-540-22913-2

Vol. 163. A. Ghosh, L.C. Jain (Eds.)
Evolutionary Computation in Data Mining,
2005
ISBN 3-540-22370-3

Vol. 164. M. Nikravesh, L.A. Zadeh,
J. Kacprzyk (Eds.)
*Soft Computing for Information Prodessing
and Analysis*, 2005
ISBN 3-540-22930-2

Armando Freitas Rocha
Eduardo Massad
Alfredo Pereira Jr.

The Brain:
Fuzzy Arithmetic to
Quantum Computing

Prof. Armando Freitas Rocha
Rue Maria Inez Carleti 26
13201-813 Juandiaí
Brazil

Prof. Eduardo Massad
University of São Paulo
Av. Dr. Arnaldo 455
São Paulo, SP. 01246-903
Brazil
E-mail: edmassad@usp.br

Dr. Alfredo Pereira Jr.
UNESP
Department of Education
18618-000 Botucatu
Brazil

ISSN 1434-9922
ISBN 3-540-21858-0 Springer Berlin Heidelberg New York

Library of Congress Control Number: 2004111138
This work is subject to copyright. All rights are reserved, whether the whole or part of the material is concerned, specifically the rights of translation, reprinting, reuse of illustrations, recitations, broadcasting, reproduction on microfilm or in any other way, and storage in data banks. Duplication of this publication or parts thereof is permitted only under the provisions of the German copyright Law of September 9, 1965, in its current version, and permission for use must always be obtained from Springer-Verlag. Violations are liable to prosecution under the German Copyright Law.

Springer is a part of Springer Science+Business Media
springeronline.com

© Springer-Verlag Berlin Heidelberg 2005
Printed in Germany

The use of general descriptive names, registered names trademarks, etc. in this publication does not imply, even in the absence of a specific statement, that such names are exempt from the relevant protective laws and regulations and therefore free for general use.

Typesetting: data delivered by authors
Cover design: E. Kirchner, Springer-Verlag, Heidelberg
Printed on acid free paper   62/3020/M - 5 4 3 2 1 0

Table of Contents

Introduction 1

1 Quantification and Calculation in Nature 7
   1.1 Why and How has Arithmetic Cognition Evolved? 7
   1.2 The Numerical Competence of Animals............................ 9
   1.3 The Numerical Competence of Human Infants 13
   1.4 A Brief Account from Neuroscience 16
   1.5 Stand Up and Count (and Get Smarter!)......................... 18

2 The Cells of the Brain 21
   2.1 Molecular Neurobiology.. 21
   2.2 Fuzzy Formal Languages... 27
   2.3 Ambiguity.. 30
   2.4 The Hierarchy of Fuzzy Grammars 36
   2.5 Fuzzy Languages, Distributed Processing and Biological Diversity 41
   2.6 Knowledge Adaptation and Evolution............................ 43
   2.7 Self-Controlled Expression.. 46
   2.8 Speeding up Brain Processing 47
   2.9 The Chemical Talk at the Synapse 50
   2.10 Summary.. 52

3 Brain: A Distributed Intelligent Processing System 53
   3.1 Distributed Intelligent Processing Systems 53
   3.2 Distributed Processed Languages 55
   3.3 The Nervous System... 59
   3.4 The Brain 60
   3.5 Brain Communication Channels..................................... 64
   3.6 The Basics of DIPS Learning 66
   3.7 Evolutionary Learning 70
   3.8 Summary.. 77

4 Neural Computational Mechanisms Supporting Cognitive Processes 79
   4.1 Basic Concepts of Quantum Computation..................... 79

4.2 Cellular Processing 82
4.3 Current Physical Implementations of Quantum Computers 86
4.4 The Dendritic Spine as a Quantum Computing Device............. 89

5 The Brain and Quantum Computation..93
5.1 The Dendritic Spine as an Ion Trap Quantum Computing Device. 93
5.2 The Deutch-Josza Algorithm 96
5.3 The Quantum Cortical Recognition Device................................ 100
   5.3.1 The Model.. 100
   5.3.2 Learning 104
   5.3.3. Recognizing Faces.. 105
5.4 Fuzzy Logic and Conflict in $O(G|H,S_b)$ 106
5.5 Recurrent Architecture Generates Entanglement Supporting
Consciousness... 109
5.6 Superdense Codes.. 113
5.7 Quantum Computing and Working Memory 114

6 Memetics and Cognitive Mathematics 117
6.1 Memes.. 117
6.2 The Formal Meme 119
6.3 Improving Meme Spread... 120
6.4 Memetic Channels and the Brain.. 122
6.5 The Evolution of our Mathematical Competence....................... 123
6.6 How Memes Spread.. 126
6.7 How the Number Meme Spread... 131
6.8 Future Perspectives in the Memetics of Cognitive Mathematics . 132

7 Modeling of Arithmetic Reasoning 135
7.1 Counting 135
7.2 A Model for Number Sense.. 136
   7.2.1 Identifying numerosities 137
   7.2.2 Quantification trajectory control....................................... 140
7.3 The Crisp Numbers... 141
7.5 Doing Arithmetic 145
7.6 The Evolution of Arithmetic Knowledge 149
7.7 Representation of Other Abstract Entities................................. 151

8 Brain Maps of Arithmetic Processes in Children and Adults........153
8.1 The Study of the Mathematical Brain 153
8.2 The Technique.. 155
   8.2.1 Theory.. 155
   8.2.2 The Methods 157

    8.3 Agent Commitment Experimentally Measured.......................... 162
        8.3.1 Males are Faster than Females in Arithmetic Calculations ... 164
        8.3.2 Size Number Effect... 169
    8.4 The Distributed Mathematical Brain 170
    8.5 Building the Distributed Mathematical Brain 175
    8.6 Conclusion 177

9 Arithmetic Learning Capability in Congenitally Injured Brains ..179
    9.1 Dyscalculia... 179
    9.2 Damaged Brains 180
    9.3 Neural Plasticity 191
    9.4 Developmental Dyscalculia 192
    9.5 Conclusion 197

10 Learning Arithmetic: Why So Difficult?...199
    10.1 The Nature of Arithmetic Knowledge 199
    10.2 The Invention of Crisp Numbers 202
    10.3 Learning by Observing 203
    10.4 Arithmetic Meme Diffusion in School..................................... 206
    10.5 Evolving Arithmetic Knowledge in the School........................ 209

References.. 213

Index.. 225

Introduction

We could start writing this book by saying, with several other authors, that the brain is the most powerful and complex information processing device known, whether naturally developed or created artificially. Although we fully agree with this statement, in doing so we would be misleading the reader, in the sense that the present book basically aims to formalize the knowledge concerning brain physiology accumulated over the past few decades. Instead of merely describing the complexity of the cerebral structure or presenting a collection of commentaries and reviews of interesting experimental results, we take into account novel achievements in quantum information and quantum computation, and avail ourselves of recently developed mathematical tools.

Neuroscience was born in the 19th century with the works of Paul Brocca. However, this fledgling field experienced a boom only in recent times, following the development of powerful non-invasive techniques for probing the neural circuitry supporting the complex cognitive functions of the human brain. Although sophisticated mathematical models and physical theories are the basic tools behind the conceptual foundations and analytical implementation of these modern techniques, to the best of our knowledge no effort was made to formalize the actual knowledge about brain function into a coherent theoretical framework incorporating the recent developments in mathematical and physical science. Addressing this lack was our first motivation in writing this book. The adequacy of Fuzzy Sets and Fuzzy Logic for such a purpose was realized by us in the pioneering days when these powerful ideas were first introduced by Prof. Lotfy Zadeh. Since the very beginning we chose Fuzzy Formal Language theory (FFL) as the fundamental formalism to implement this task. According to modern neuroscience, chains of chemical transactions, called Signal Transduction Pathways (**stp**), are the basis for explaining brain function. In our approach, fuzzy sentences, supported by a fuzzy grammar **G,** provide formal descriptions of biochemical transactions, or stps, within and between neurons. Soft Computing is characterized as the set of FFL techniques used by the brain to solve both control and cognitive tasks.

The famously strange properties of quantum mechanics have been proposed to explain complex cerebral functions since Penrose proposed that consciousness is the result of quantum transactions. But consideration of these properties is also giving rise to new theoretical computational techniques for qualitatively and quantitatively enhancing information processing, techniques which are fast becoming a practical reality with the recent experiments in quantum computation employing both Ion Trap and MNR procedures. Thus, quantum properties of ions or molecules are used to create quantum gates by way of unitary transformations that form the logical basis of quantum computation. Nonlocal properties theoretically allow for massively parallel processing that promises to greatly enhance the computational power of distributed processing systems like the brain. Moreover, recent experimental data from neuroscience support the hypothesis that certain neural structures, like the dendritic spine, are specialized to function as Quantum Computing Devices, taking advantage of the quantum mechanical properties of specific stps. These kinds of considerations inform our view of the brain herein, which we describe as a Quantum Distributed Intelligent Processing System.

Neuroscience is demonstrating the existence of inherited or innate cognitive modules for language, arithmetic, biology, physics, and music. This knowledge is expected to dramatically influence the way we approach modern education in the strongly competitive technological society of the present day. This is our second main motivation in writing this book. It is a well-established fact that current methods for teaching arithmetic are influenced by our cultural belief that numbers are abstract entities living a life of their own and that numerical knowledge distinguishes man from the animals. This is in striking contrast with the experimental evidence of the existence of number sense in animals and in the human infant. But if man is phylogenetically endowed with arithmetic neural circuits, why is learning (and teaching) mathematics so stressful?

Recent experimental results demonstrate that learning by observing is a powerful way whereby animals may quickly acquire the competence to solve many kinds of problems, by simply watching another animal discovering how to arrive at solutions to these challenges. However, as these same experiments also show, observing an incomplete portion of the entire sequence of discovery is insufficient for this type of learning; moreover, partial observation also precludes the animal from discovering the best solution on its own. Perhaps because traditional mathematical teaching does not take into account the innate human capability for arithmetic, our teachers demonstrate a suboptimal means of doing arithmetic. This not only causes initial distress to the pupil, but perhaps may lead to a permanent dislike or fear of mathematics. If this assumption turns out to be correct, then actual knowledge about the arithmetic brain, as formalized in this book, may be used to plan a better and happier way of teaching not only arithmetic but mathematics as a whole.

For the above reasons, we believe that neuroscientists will profit from taking notice of our proposals in developing powerful neural models to guide their experiments and to better interpret their results. By doing so, they will describe their discoveries in a more comprehensible way, and for a broader audience. Mathematicians could discover an entire new field, not only for applications, but also for new developments in their own area, which could feedback in the near future to a new wave of discoveries about brain function. The emerging field of quantum computation will profit from understanding that quantum information processing is a technique discovered by natural selection some millions of year ago. This, we suppose, makes a strong case for the hope of creating artificial quantum computers in the near future. The understanding of the brain as a quantum processing intelligent system, together with the creation of artificial quantum devices, will also expand the field for experimental physics to test the strange properties of quantum mechanics. This makes the present work interesting to the theoretical physicist.

But above all, we hope teachers and all other related professionals may profit from understanding how the brain they have to teach operates, in order to easy the thorny pathway that youngsters have to travel throughout their schools years. Most especially, we hope that understanding the functions and plasticity of arithmetic neural circuits may help educators make learning arithmetic less stressful.

To accomplish such purposes, this book is organized as follows.

Chapter 1 is devoted to a review of the literature, showing that:
1. several species of animals identify and manipulate the cardinality of sets of elements important to their survival;
2. the human infant has a capability for recognizing and operating on small quantities;
3. the brain is organized to handle quantities and to have a number sense.

The opening chapter also introduces the hypothesis that human arithmetic capability is the result of the mutual influence of genetically inherited and culturally transmitted information.

Chapter 2 introduces the basic concepts of Fuzzy Formal Languages and Signal Transduction Pathways, and how the physiology of the neuron may be understood, taking into account these concepts. In this context, a grammatical hierarchy is introduced and some basic theorems are provided. The synaptic transactions are described as processes supported by a Self-Controlled Grammar **G**, and become dependent on both the quantity and the quality of the chemicals involved. The expression of the languages

L(G) supported by **G** is subject to constraints imposed by limited resources available in the external environment. The neuron is therefore formalized as a symbolic processing machine, in contrast to the numerical approach of classic neural network theory. It is also shown that the McCulloch-Pitts neuron is a special case of our model, if the number of sentences is considered, and not their symbolic structure.

Chapter 3 discusses the brain as a Distributed Intelligent Processing System (**DIPS**), whose agents are specialized in handling different subsets of the language L(G) supported by the grammar G. The ontogeny of the brain is formalized by a specialized subset of L(G) obeying the rules discovered by biology for the functioning of a set of genes called *homeobox* genes. Reasoning is proposed to be the result of transactions among a complex set of agents. Learning is assumed to obey a set of rules controlling both the creation of new agents by specifying expressible subsets of L(G), as well as communication among agents, by setting the amount of resources available for synaptic transactions. Learning, in such a context, is viewed as an evolutionary process, by means of which an initial set of models is set to evolve under the influence of the (material and cultural) resources provided by the external environment.

Chapter 4 introduces the basics of quantum computation and quantum information, and discusses how molecular transactions at the cellular level may implement such concepts. We propose that the brain uses a quantum computational strategy, allowing microstate entanglement of several spatially distributed processing units, thereby providing a supplementary communication channel to overcome possible mismatches. This channel also provides instantaneous binding of the informational content being processed in such units.

Chapter 5 describes dendritic spines (**DS**) as specialized synaptic structures for trapping Ca^{2+} ions and as very plastic structures involved in both rapid (imprinting) and slow (reaction to environmental changes) learning. DS are assumed to have quantum computing capability and some calculations are presented to support such a claim. A DS model of the Deutsch-Josza algorithm is presented and used to show how quantum computation may be used by the brain for pattern-recognition purposes.

Chapter 6 introduces the logic and basic assumptions of memetics, presenting its definition and proposing memetics as a potential answer to the problem of brain evolution. The human capacity for dealing with complex mathematical processes is proposed to be the result of a meme-gene interaction. A simulation study illustrates how knowledge spreads in a small community, when information was provided by means of "broadcasting" and "mail" systems. In this context, evolution of mathematical knowledge in human culture is proposed to be the result of a meme-gene coevolution of brain size and complexity.

Chapter 7 proposes a new class of fuzzy numbers, called K Fuzzy Numbers (**KFN**), to model arithmetical knowledge in the brain of animals and humans. Three populations of agents - Controllers; Accumulators and Quantifiers - are assumed to handle **KFN**. Also, the evolution of the accumulating function (changing from a monotonic to a periodic function), and of the classifying function, are proposed as the means whereby **KFN** evolved to create Crisp Base Numbers (**CBN**) like our decimal system. This may explain the innate human capacity to create complex number theories and arithmetic.

Chapter 8 describes the experimental results on performance in solving arithmetic calculations by adults and children while having their EEG activity recorded. The experimental data show:

1. a clear-cut distinction between genders, males being faster than females in providing an equally correct answer;
2. a quick learning characterized by a calculation time decrease, which is dependent on the order of problem presentation: and
3. a number size dependence of the calculation time that differed for children and adults.

These data are interpreted in the framework of the model proposed in Chap. 7.

Chapter 9 deals with the capacity for learning arithmetic among brain damaged children with low and normal IQs. The results clearly demonstrate that:

1. widely distributed bilateral parietal lesions reduce the children's arithmetic learning capability;
2. left frontal lesions may dissociate the capability of handling and operating ordinal and cardinal numbers; and
3. brain plasticity allows children to overcome most of their arithmetic learning problems.

Also, the results corroborate the conclusions about the organization of the arithmetic neural circuits discussed in chapter 8.

Finally, chapter 10 assumes that learning arithmetic implies the development of inherited **CBN** circuits under the guidance of teachers. This process is proposed to follow an orderly pathway that must begin with the construction of many different circuits in distinct cerebral areas, a process triggered by questions posed in a set of problems of increasing complexity.

Based on experimental results from the literature on "learning by observing," we argue that, when the evolutionary nature of mathematical learning is not taken into account, the development of arithmetic knowledge is seriously compromised, rendering instruction less effective.

This book owes a lot to the many children and adults who volunteered in the experiments we conducted, wherein we probed many human cognitive functions. We thank them very much for the brain time they shared with us. We are also indebted to the parents of these normal and disabled children, who understood that, through research, their youngsters pave the way for a better education for succeeding generations. We must also acknowledge the participation of many teachers and other persons involved in the instruction of these of children. We would like to stress the special participation of people from Colégio Clip – Guarulhos and APAE-Jundiai, where we collected most of the experimental data discussed in Chaps. 8 and 9.

The authors are also indebted to several colleagues and friends who, through discussions and invaluable comments, greatly contributed to the final result. Among them we especially acknowledge Francisco Antonio Bezerra Coutinho, Fernando Gomide, Witold Pedrycz and Janusz Kapcrzyk. We are also grateful for the technical support provided by Brian J. Flanagan, Marcos Paulo Rebelo, Mateus Fuini, Ednilson Rodela e Fabio Luiz Picceli Luchini. But above all, we are in very special debt to Cassia Medea, who provided us with the illustrations used throughout the book. Other friends, from the Discipline of Medical Informatics and LIM01/HCFMUSP of the University of São Paulo, provided the right environment, which greatly stimulated the production of this book.

1 Quantification and Calculation in Nature

Several species of animals identify and manipulate the cardinality of sets of elements important to their survival. The human infant has the capability to recognize and manipulate small quantities. These facts will be used to support the hypothesis that specific neural circuits have evolved in nature to mediate these capabilities and that these circuits constitute the foundations for the further development of mathematics by modern man.

1.1 Why and How has Arithmetic Cognition Evolved?

When the English philosopher John Locke got in touch with the Indians from the fierce tribe of the Tupinambás (who lived in what is now the State of São Paulo, Brazil, and were well known for their ferocity and cannibalism), he not only escaped violent death but even noted that their language lacked names for numbers above five (Fig. 1.1). When the Tupinambás went beyond five they simply showed their fingers and the fingers of others (Butterworth, 1999). As a matter of fact, the Tupi language has names for the first four numbers, whether used as cardinals or ordinals (Navarro, 1998). After four, quantities were referred to by saying "nhã" and showing the corresponding number of fingers. The number ten was referred to as "my hands" and twenty by "my hands and feet."

The fact that these Native Americans never developed a full language for numbers reflects, first, the sheer lack of necessity – trade was not their cup of tea. In the first place, quantities above five in general did not deserve special status and were denoted by "many." Secondly, our innate counting capability is similar to that enjoyed by several animals (e.g., Dehaene, 1997; Gallistel and Gelman, 2000; Shettleworth, 1998).

We are, therefore, born with a capacity to enumerate objects, which is strictly limited to some items, and we share this genetically determined characteristic with more "primitive" animals. Several empirical studies have demonstrated that the counting capacity is innate in rats and pigeons, and there is even an observable and remarkable competence in human infants for simple arithmetic (Wynn, 1998). This genetically determined arithmetical competence has been studied by several authors since the beginning of the 20th century.

Fig. 1.1. The Tupinambá Number System: *How far is this place? Five moons walking.*

Some interesting experiments in which human subjects are asked to enumerate objects have shown that enumerating a collection of items is fast when there are one, two or three items, but starts slowing drastically beyond four. In addition, errors begin to accumulate at the same point (Dehane, 1991; Fayol, 1996; Gallistel and Gelman, 1991, 2000). It was hypothesized that "mental operations with verbally or visually presented digits depends on a mapping to mental magnitudes that seems to obey Weber's law," which proposes a logarithmic rather than a linear relation between "numbers" and magnitudes (Dehaene, 2003; Dehaene et al. 1998; Gallistel and Gelman, 1991, 2000 ; McCloskey et al. 1991; Nieder et al. 2002). To better explain why numbers below five are quickly identified at the almost the same speed, it was proposed that counting in this condition is performed in blocks, a process called *subtizing* in the literature (Butterworth, 1999, Dehane, 1997; Fink et al. 2001). It takes about five to sixtenth of a second to identify a set of three dots, about the time it takes to read a word aloud or to identify a familiar face, and this time slowly increases from 1 to 3 dots.

How and why have arithmetic cognitive abilities evolved? It is tempting to answer this question with the obvious notion that the larger the brain the greater its owner's capacity to adapt and survive in aggressive and/or rapidly changing environments. But one could then argue that, in addition to the size of the brain, its configuration and neuronal specialization have roles to play. But is there a cerebral region responsible for mathematical thinking? The first experiments, carried out in the early 1980s, demonstrated higher cerebral activity in the inferior parietal cortex as well as multiple regions of the prefrontal cortex in the course of numerical performance tasks. Recent experiments with functional magnetic resonance demonstrated that several other cerebral areas are activated during mental calculations (Butterworth, 1999, Dehane et al. 1998; Göbel et al. 2001; Nieder et al. 2002; Sawamura et al. 2002, Zorzi et al. 2002). It is now accepted that the inferior parietal region is important for the quantification of cardinalities, and the representation of relative number magnitudes. The extended prefrontal cortex is, in turn, responsible for sequential ordering of operations, control over their execution, error correction, inhibition of verbal responses, etc. Therefore, arithmetic neural circuits are now believed to be widely distributed over several brain areas.

1.2 The Numerical Competence of Animals

There is a good deal of evidence that many animals, from pigeons to rats to chimpanzees, have a certain, if limited, numerical competence. Apart from some famous hoaxes, like the well-known case of 'clever Hans' (the horse that was supposedly able to perform some calculations), many experiments have demonstrated that there is an innate numerical competence, which evolved along the phylogenetic scale, increasing dramatically with human beings. Although most experiments with animals required extensive training, numerically relevant behavior has also been observed in the wild.

In the 1950s and 1960s, some animal psychologists from Columbia University (Mechner, 1958 and Dehaene, 1997) performed a series of experiments with starving rats, which were then induced to press a lever a certain number of times to get a given amount of food (Fig. 1.2). They demonstrated that the rats not only learned to press the level to receive the food but also that their responses were close to the expected ones. The differences between the expected and the actual number of lever pressings increased with the size of the number. The variance of the distribution of the actual number of lever pressings also increased with the size of the number of lever pressings (Fig. 1.3). Other recent experiments resulted in similar results (Platt and Johnson, 1971; Gallistel and Gelman, 2000).

Fig. 1.2.a The Rat Number System: *press the lever x times to get food (adapted from Mechner experiments described in Dehaene, 1997)*

These experiments demonstrated that the numerical competence of small-brained animals is surprisingly good but limited to small quantities. As shown in Table 1.1, the distances between the desired and the actual number of presses, as well as the width of the distributions, increased in proportion to their mode. This trial-to-trial variability in the accuracy with which the animals approximated target numbers was proportional to the magnitude of the target, even for a number as small as four (Gallistel and Gelman 2000).

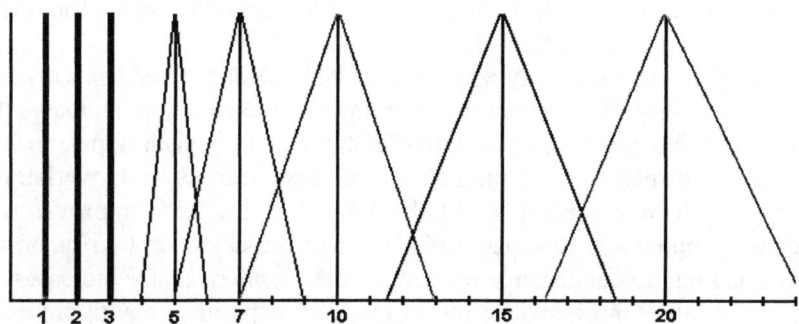

Fig. 1.2.b Variance of Lever Pressing

A fundamental question in cognitive science is whether animals can represent numerosity and use numerical representations computationally. In a recent experiment, Brannon and Terrace (1998) showed that Rhesus monkeys represent the numerosity of visual stimuli and detect their ordinal disparity.

Table 1.1. Lever pressing statistics

D #: pressings	A #: pressings	D – A	Variance
4	4.5	0.5	1.25
8	9.0	1.0	6.25
12	13.5	1.5	12.25
16	18.0	2.0	16.00

D#: Desired number of lever pressings
A#: Actual number of lever pressings
V: Variance of distribution in actual number of lever pressings

Two monkeys were first trained to respond to exemplars of the numerosities 1–4 in ascending numerical order. As a control for nonnumerical cues, exemplars were varied in size, shape and color. The monkeys were later tested without reward, on their ability to order stimulus pairs composed of the novel numerosities 5–9. Both monkeys responded to ascending order in the novel numerosities, demonstrating that Rhesus monkeys represent the numbers 1–9 on an ordinal scale.

In the next year, Carpenter et al.(1999) reported neurons located in the monkey motor cortex that are specifically sensitive to the order of presentation of visual stimuli, providing the physiological substrate for the results of Brannon and Terrace (1998).

In another experiment, Nieder et al. (2002) demonstrated the existence of a specific set of neurons in the lateral prefrontal cortex of monkeys that were tuned for quantity, irrespective of the exact physical appearance of the training displays. The tuning curves of those neurons form overlapping filters, which may explain why behavioral discrimination improves with increasing numerical distance and why discrimination of two quantities with equal numerical distance worsens as their numerical size increases.

But the more impressive results come from experiments with chimpanzees, particularly in the work by Woodruff and Premack (1981). These experiments demonstrated that chimpanzees are able to perform rather sophisticated abstract computation. They are able, among other things, to conclude that one-quarter of a pie is to a whole pie as one-quarter of a glass of milk is to a full glass of milk. The authors showed, in addition, that chimpanzees could even mentally combine two fractions. If they had to add one-quarter apple 'plus' one-half glass of milk, and they had to choose between a full disc or three-quarter disc, the animals correctly chose the latter, and the result was statistically significant.

Chimpanzees trained with Arabic digits can even identify two Arabic digits (such as 2 and 3) and point to their sum (as 5) amidst other Arabic digits (Dehaene et al. 1998). In addition to an impressive numerical competence, chimpanzees also display a fairly good numerical memory span. In a recent experiment, Kawi and Matsuzawa (2000) tested the memory span and other numerical skills of a female chimpanzee called Ai. These researchers designed their experiment based on the fact that humans can easily memorize strings of codes such as phone numbers and postal codes if they consist of up to seven items. They therefore tried to determine the equivalent limit in chimpanzees. Their design demonstrated that chance levels with three, four and five items were 50, 13, and 6%. The chimpanzees scored more than 90% with four items and about 65% with five items, significantly above the chance in each case. The results indicate that chimpanzees can remember the sequence of at least five numbers, the same as preschool children.

Finally, let us briefly review the most recent research on the numerical competence of primates, specifically, the search for a 'cerebral seat' of numerical representation in monkeys. In an elegant experiment, Sawamura et al. (2002) showed that the anterior part of the parietal association area in the cerebral cortex is active in primates performing numerically based behavioral tasks. The authors required monkeys to select and perform movement A five times, switch to movement B for five repetitions, and return to A, in a cyclical fashion. Cellular activity in the superior parietal lobe reflected the number of self-movement executions. For the most part, the number-selective activity was also specific for the type of movement. The authors reported that this type of numerical representation of self-action was seen less often in the inferior parietal lobe, and rarely in the primary somato-sensory cortex. According to the authors, such activity in the superior parietal lobe is useful for processing numerical information, which is necessary to provide a foundation for the forthcoming motor selection.

1.3 The Numerical Competence of Human Infants

It is now a well-established fact that in the first few months of life, human infants can enumerate sets of entities and perform numerical computations. The capacity to represent approximate numerosity, found in adult animals and humans, develops in human infants prior to language and symbolic counting (Wynn, 1998; Xu and Spelke, 2000). However, it remains an open question in cognitive sciences, philosophy and psychology as to how we acquire our knowledge of *number*.

One prevalent view is that the numerical abilities of humans arise from general cognitive capacities not specific to number. However, a growing number of researchers have been arguing that the current body of data supports the thesis that humans possess a specialized mental mechanism for number, one which we share with other species and which has evolved through natural selection (Wynn, 1998). Recent experiments demonstrate that infants represent cardinal values of small sets of objects (Feigenson et al., 2002). In addition, as early as six months of age, infants discriminate between large sets of objects on the basis of numerosities when other extraneous variables are controlled, provided that the sets to be discriminated differ by a large ratio (8 vs.16, e.g., but not 8 vs.12) (Xu and Spelke, 2000).

The classic experiments to assess numerosity awareness in children deal with time spent looking at displayed objects (Starkey and Cooper, 1980 and Strauss and Curtis, 1981). This parameter has demonstrated good sensitivity in quantifying children's interest, and shows that infants can discriminate between different small numbers of entities (Wynn, 1998). Thus, when repeatedly presented with displays of a given number of visual objects, infants become bored and spend less time looking at the displays. If the number of objects displayed is changed, children regain interest and look for longer periods.

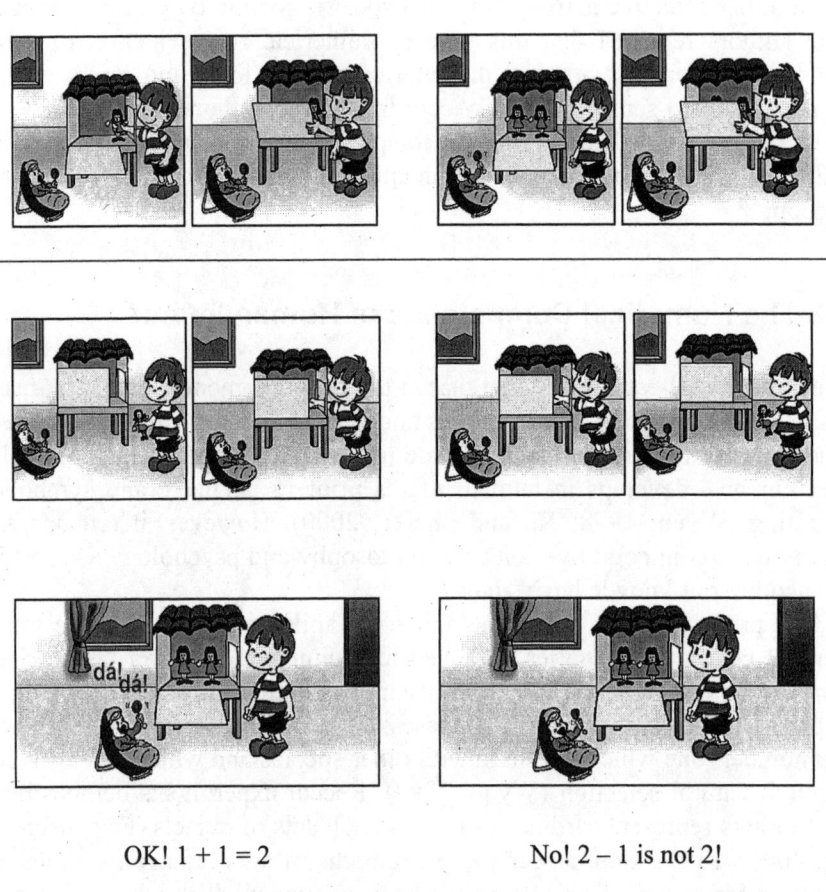

OK! 1 + 1 = 2   No! 2 − 1 is not 2!

Fig. 1.3 Studying the Baby Number System: *the baby looks longer when 2 − 1 = 2 than when 1 + 1 = 2*

In addition to number sensitivity, small children are able to engage in numerical computation. When presented with a 1 + 1 operation and a 2 − 1 operation in looking time experiments (Wynn, 1996), infants showed significant differences in their looking time with respect to results representing correct and incorrect answers (Fig. 1.3).

The *"greater than"* and *"lesser than"* relations between numbers constitute another essential component of our number system. In a recent experiment, Brannon (2002) showed that 11-month old infants successfully discriminated, whereas 9-month old infants failed to discriminate, between sequences of descending numerosities from sequences of increasing numerosities. This suggests that 11-month old infants can appreciate the "greater than" and "lesser than" relationships between numerical values, a characteristic that develops after nine months of age.

Wynn (1998) proposed that there exists a mental mechanism, dedicated to representing and reasoning about number, that constitutes part of the inherent structure of the human mind. In addition, as we humans share this numerical discrimination with a large set of warm-blooded vertebrates, it is reasonable to conclude that such a characteristic has a strong adaptation value, and has therefore quite likely evolved by natural selection — a conclusion borne out by many similarities between infants and animals' numerical abilities: animals as well as human infants enumerate a wide range of entities, including moving or stationary objects and events, presented simultaneously or sequentially (Davis and Perusse, 1988; Gallistel, 1990; Wynn, 1998).

A compelling piece of evidence regarding the nature and development of numerical knowledge is the historical development and ontological discussion of the concept of *zero*. The so-called place-value system of numeral notation took over 1500 years to introduce the notation for zero (Ifrah, 1985, Joseph, 1990). It is arguably more than a coincidence that our individual difficulties with the numeral zero is mirrored in the historical development of that concept. The individual understanding of zero does not follow the same pattern of development as children's understanding of positive integer values (Wynn, 1998). It normally takes a long time before infants appreciate that zero is a numerical value, even when children have already learned what the word zero and its corresponding Arabic numeral symbol stand for. In another example provided by Wynn (1998), preschool children will name "one" as the smallest number, even when they know that the word zero applies to "no items."

In conclusion, there is strong evidence showing that small children can enumerate different kinds of entities, and can compute the numerical outcomes of operations on small numbers of entities. Also, accumulated experimental findings support the existence of a dedicated brain mechanism specific to numerosities, which serves as a foundational core of numerical knowledge for our sophisticated mathematical abilities.

1.4 A Brief Account from Neuroscience

The famous neurologist Oliver Sacks, in his book *The Man Who Mistook His Wife for a Hat* (1985), wrote that 'deficit' is the favorite word of neurologists. By this he meant the deficiency and loss of capacity so pervasive in neuropathologies. Indeed, several deficiencies are directly attributable to lesions of more or less specific areas of the nervous system, and the science of neurology owes a great deal to such deficits, or 'experiments of nature' in the words of Dehaene (1997).

The first scientist to associate deficits with function was Paul Brocca, the famous French physician, who in the 19th century correlated language impairment with lesions of a very specific area in the frontal lobe, later named *Brocca's* area after him. Gertzmann was the first to associate arithmetical cognitive disability with brain damage, when he described the case of a 52-year-old woman, who was admitted at the Wiener Psychiastriche Klinik, complaining of difficulties with memory and writing ability. Neurological examination showed calculation impairment, writing disability, lack of recognition and orientation of her own body, and incorrect selection and orientation of individual fingers or hands. This latter deficit is called finger agnosia. This clinical picture was shown to be the result of a very specific left parietal lobe lesion compromising finger movements, writing and arithmetic. This complex of clinical manifestations, due to a brain lesion, is now called *Gertzmann Syndrome* (Gertzmann, 1924, see also a recent case report by Mayer et al. 1999). The current understanding of the distribution of numerical processing in the brain is largely due to studies regarding *dissociation* of arithmetical capabilities resulting from lesions on many different areas.

Dissociation is an important phenomenon in cognitive neuroscience, and refers to the fact that, following cerebral damage, one domain of competence becomes inaccessible while another remains largely intact. So, in some patients, severe difficulties in reading Arabic numerals have been reported. In other cases patients can display double dissociation. For instance, one patient had the grammatical structure of numerals intact, whereas for a second patient this faculty had deteriorated; in contrast, the selection of individual words was deficient in the first and intact in the second. As a result, the first patient often replaced one numeral with another but he never erred in the decomposition of a number. For instance, he was able to read 681 as "six hundred *fifty*-one", that is, the structure of the string is correct except for the substitution of *fifty* for *eighty*. The second patient, in contrast, never took 1 for "two," like the first, but misread 7,900 as "seven thousand ninety". The conclusion of these clinical cases is that perhaps some of the cerebral regions engaged in reading Arabic numerals aloud contribute more heavily to number grammar, while others are more concerned with accessing a mental lexicon for individual numeral words (Dehaene, 1997).

The recent development of non-invasive techniques for the study of the living human brain is bringing new light to bear on the neurosciences. Brain mapping machines, like MRI (Magnetic Resonance Imaging), fMRI (functional MRI), SPECT (Single-Photon Emission Computerized Tomography), PET (Positron Emission Tomography), MEG (Magneto-Encephalography), and EEG (Electro-Encephalography), all heavily dependent on computer power, are changing the way scientists explore the function of the brain. These new techniques are disclosing fundamental functional contributions of frontal and parietal neurons to numerical processing, as well as the existence of different neural circuits to deal with both approximate and precise numerical reasoning.

But perhaps the greatest contributions of these techniques for the questions at hand consist in facilitating the study of the plasticity of neuronal functioning in respect of numerical competence. The marriage of brain-lesion studies and non-invasive techniques is showing that, while congenital lesions may be devastating in brain tissue loss, these kinds of damage do not restrain children from building up new circuits at new places in order to acquire numerical competence as complex as that attained by the normal brain (e.g. Rocha et al. 2003b). These authors have followed up the cognitive evolution of three congenitally brain damaged children who experienced delayed linguistic and arithmetical skills acquisition, described in details in Chapter IX.

One of the children lost most of her left-frontal lobe, spoke her first words at the age of five, and began to read and write simple words, as well as master quantities above five, at the age of eleven. At the age of thirteen, she started to perform very well at addition and subtraction — after she invented a way to cope with her dissociation; she was able to quantify but she had problems in (orally/or graphically) naming these quantities.

Another girl had her right parietal lobe damaged, and acquired language at the normal age, but experienced early difficulties with arithmetic. At the age of sixteen, she began to master complex calculations with up to three-digit numbers.

The third case is that of a boy who lost his left-parietal lobe, started to use his first phrases at the age of five years, began to understand quantities above five at the age of seven, and to perform very simple calculations (one-digit summation and subtraction) at the age of ten. At the age of fourteen he started to master addition and subtraction up to three digits and multiplication of up to two digits. Brain mapping clearly showed a reorganization of neural circuits paralleling such developmental achievements.

Contrasting with these examples of severe brain damage and reasonable performance, some lesions that go clinically undetected (even by the majority of those techniques described above) may drastically change the academic life of many children, as in the case of developmental dyscalculia. Around 5% of school children experience minor brain cell deaths that results in an abnormal EEG pattern (called Intermittent Rhythmic Delta Activity), associated with one or more of the following symptoms: hyperactivity, attention deficit, dyscalculia and dyslexia.

In conclusion, data gathered by the neurosciences appears to indicate that mathematical reasoning is deeply seated in widely distributed and very plastic neural circuits.

1.5 Stand Up and Count (and Get Smarter!)

It is difficult to be precise as to when our forebears evolved from the chimpanzee-like, knuckle-based gait to the upright position and bipedal gait of modern humans. This important change in the way our ancestors walked had tremendous consequences for our brain development.

Recent molecular biology dating, based on DNA comparison and assuming a constant mutation clock, points to a time when our species first diverged from our cousin chimpanzees between 5 and 7 million years ago (Marks, 2002). However, as we share about 98.5% of our DNA with the chimpanzees, it is difficult to account for the differences between us and them, based only on the meager 1.5% of non-paring genetic material. We will therefore concentrate on the fossil record, which points to important skeletal transformations which occurred when the first members of our lineage dropped down from the African trees, in order to try to understand these differences.

Our genus is denominated *Homo* and it is now widely accepted that we evolved from a line of other primates, called *hominids*, from whom we split some 2.5 million years ago. Some twenty different species of hominids are listed in the current literature (Tattersall and Schwartz, 2000). What is important for our subject, however, are the implications of bipedalism on brain development, including the consequent correlation between bipedalism and the numerical competence of our ancestors.

Two basic models of human development can be identified in the current literature (Tattersall and Schwartz, 2000). The first model is called "linear" and it states that the anatomical characteristics of the hominids appeared once in the phylogenetic record and all the descendants built upon the basic features in a linear fashion. The second model, called the "disordered" model, argues that hominid evolution happened through a set of evolutionary diversification events, during which the anatomical characteristics mixed up in a way we have only recently begun to comprehend.

What is important for our discussion is the very definition of hominid. According to the specialized literature, the hominids are defined by the structure of the face and teeth as well as by bipedalism. The recent discovery of a candidate for our most ancient ancestor stirred debate within the anthropological community on the definition of hominid; this was *Sahelanthropus tchadensis* (Brunet, 2002), with an estimated age of 6 or 7 million years. This was due to the fact that only the facial bones and some teeth were found; the absence of post-cranial fossil remains renders the determination of its gait difficult, with the consequence that the determination of that species as our forebear is highly controversial (Wolpoff et al. 2002).

In any case, when the first hominid dropped from the trees and started to forage the African savanna, some important changes in its brain began a long series of adaptations that culminated with our own brain. The most important aspect of this primitive bipedalism was the liberation of the hands, which, along with better gait control, demanded an improvement in the neuro-motor systems. This process was perhaps the most important event in human evolution so far as brain development is concerned, and resulted in tool-making culture, some two million years ago. Also, the upright position put other strong demands on the evolution of the visual system, including better control of eye movement. Free hands and smarter eyes greatly improved the ability to correctly focus on complex collections of objects and count them. This started a process which demanded more intellectual processing, which in turn resulted in more sophisticated counting processes.

The relatively sudden explosion in brain size occurred between *Homo erectus* and *Homo heidelbergensis*, between 700 and 500 thousand years ago, when the brain almost doubled in size, from around 500 cc to 1000 cc., and coincided with the discovery of fire. It is possible that cooking preserved food for longer periods of time, which implies less frequent hunting and more socialization. This sparked the beginning of the modern human intellectual enterprise. First, the development of a lithic tool-making industry increased the human power for hunting and raising crops. Second, this development freed time for culture, initially represented by religion and burial rituals, but soon complemented by painting in order to record the history of the human activities (Fig. 1.4).

20 1 Quantification and Calculation in Nature

Fig. 1.4. Pedra Furada, 6–12 thousand years ago: all around the world, man begun to paint cave walls

This history telling required quantification of many of these activities, and primitive number representation started to appear as sequences of marks denoting, e.g., the number of hunted animals. Trade, the exchange of goods, soon developed into a common activity within and between groups of people. The subsequent increase in the complexity of trade required the corresponding development of the capacity of number processing to deal with it.

2 The Cells of the Brain

The neuron is modeled as a fuzzy formal language processing device, instead of a numerical processor as in classical neural nets. This is because modern neuroscience shows that synaptic events involve complex chains of chemical transactions triggered by the coupling of transmitters or neuromodulators to membrane receptors. These chains of chemical transactions are called Signal Transduction Pathways (**stps**) and are responsible for complex cellular processing, which may include gene reading. Assuming the neuron is a fuzzy language processor sets the background for discussing the brain as a Distributed Intelligent Processing System (**DIPS**) in the next chapter, and for understanding it as a quantum computer in Chaps. 4 and 5.

2.1 Molecular Neurobiology

Since Galvani's classical experiments, electrical membrane gradients and their variations have played an important role in the understanding of cerebral physiology. The work of Hodgkin and Huxley in the first half of the 20th century, for which they were awarded a Nobel Prize, clearly identified the main components governing the membrane's electrical behavior, and formalized this behavior as equivalent to that of a dynamic system having two stable, and one unstable, equilibrium state(s) (for further discussion of this subject, see Rocha, 1992). Another promising approach to the understanding of brain function has been molecular neurobiology. Combining measurement of brain activity and observation of concomitant behavior, in both normal and genetically modified animals, brain scientists have been able to understand the role of macromolecules in processes that support the observed behavior (e.g., Smythies, 2002; Bickle, 2003). Cognitive and affective molecular neurobiology are current research areas where biochemical properties of macromolecules, and the processes determined by their interactions, have been demonstrated to correlate with cognitive and affective functions of the brain, which putatively support the observed behavior. The construction of explanatory models based on molecular neurobiology focuses on interacting biological macromolecules forming a functional unity, the cell.

The individual cell can be considered as a cluster of signaling elements, comprising the DNA, the mRNA, the proteins produced by them, and functional ions, such as Na+, K+, and Ca++. The interaction of such signaling elements forms ordered chains of chemical transactions, called "signal transduction pathways" or stps, for short. Such stps include, therefore, signals that come from the outside (e.g., transmitters that bind to neuronal membrane receptors, hormones that come in blood flow), all signals that are operative inside the cell, and signals released by the cell to the outside (e.g., transmitters released into the synaptic space).

More importantly, we also consider ion movements across gates and generating bioelectrical processes (as in the formation of action potentials at the dendrites and soma of neurons) to be parts of stps. This is a useful way to avoid the frequent dualism of electrical and chemical explanations of cerebral processes. In fact, bioelectrical processes are composed of ionic fluxes that can be analyzed into strings of discrete signals, such as the passage of Na+ and K+ ions in and out of the neuronal membrane through specialized gates. Of course, such currents are measured in EEGs and MEGs as continuously varying electromagnetic fields. The discrete and continuous variants would be different forms of description of the same phenomenon.

Cellular specialization is defined in terms of the set of stps realized in the cell. DNA is the same in all cells of a given organism, but the protein pool (the proteome) is different. The proteome in each cell comprises thousands of different proteins, which are not identical in different tissues, organs, and systems of the body, although the amino acids present in each one are all determined by a DNA nucleotide sequence. Each pool is in fact a small subset of the total combinatorial possibilities of the DNA, whose composition depends on many factors (transcriptional and post-translational modifications driven by developmental and environmental forces, RNA splicing, etc.; see Bray et al. 2003) that direct genetic expression towards a subset of the total possibilities. Different types of neurons are characterized by the specific stps they express under the control of both genetic and environmental factors.

The gene is composed of two basic nucleotide strings (Fig. 2.1):
1. The **code string:** containing the nucleotide string encoding the amino acid sequence of a given family of proteins. The reading of the code string results in the synthesis of the mRNA, which is in charge of conveying the genetic information to the ribosomes, where polypeptide synthesis occurs. The code string is composed of exons and introns — that is to say, meaningful nucleotide sequences (exons) are mixed with

meaningless nucleotide sequences (introns) in the same gene. The introns are excised from the mRNA, after the DNA reading, by the action of specific molecules, which are able to recognize its delimiting sequences. The remaining string of exons contains the protein code. Therefore, DNA encoding is now assumed to be a non-linear process.

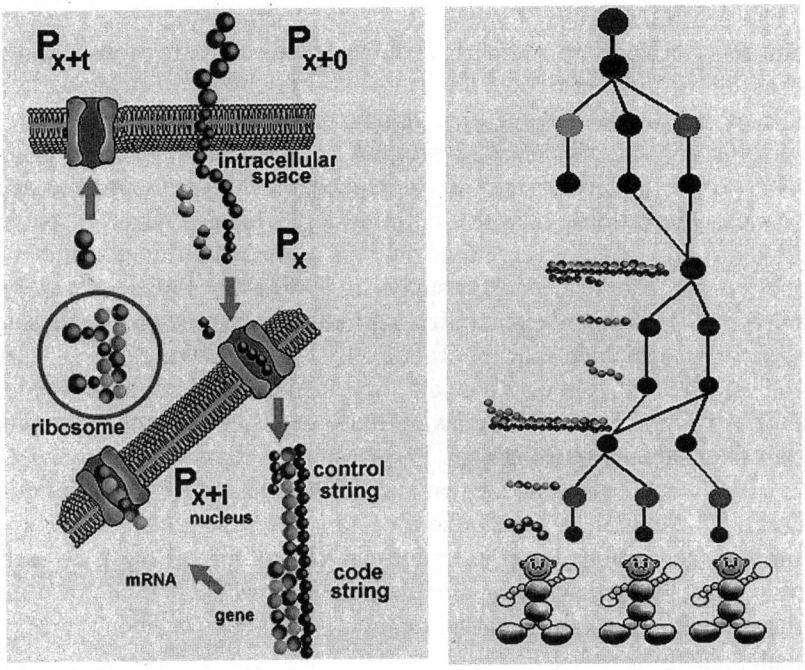

Biochemical sequences   Formal structure

Fig. 2.1. Signal Transduction Pathways (STPs): *the sentences of the computational language used by neurons*

2. The **control string**: containing the nucleotide string encoding the conditions necessary to enable and to repress the code reading. The control string is, in turn, assumed to be composed of the following substrings:
- **TATA box (promoter)**: a nucleotide sequence composed mainly of thymine and adenine, which must be activated in order to enable the code reading;

- **Inducer substrings**: the activation of these nucleotide substrings accelerates the code reading, enhancing the number of available copies of its associated mRNA, and
- **Repressor substrings**: the activation of these nucleotide substrings represses the code reading, reducing the number of available copies of its associated mRNA.

The control of DNA reading is the way nature found to control the number of proteins involved in each stp. Activation of gene reading depends on the binding of specific proteins to their control strings. The proteins are in turn specified by their genetic code strings.

In this context, the set **P** of proteins encoded by a set **D** of genes or DNA (RNA) molecules may be organized into different ordered sets of chemical interactions, each one defining a biochemical chain or pathway, called a *signal transduction pathway*. Sets of these stps are associated with functional biochemical systems supporting complex cellular activities such as cellular metabolism, excitability, reproduction, etc. Some proteins act as triggers (P_{x+0}) of these biochemical chains (Fig. 2.1), while other subsets of proteins (P_x ..., P_{x+i}) are activated at intermediary steps or as end products (P_{x+t}) of the corresponding pathway. Early genes control the expression of other downstream genes (Fig. 2.1) that specify the sets of proteins (P_{x+0}, P_{x+i}, P_{x+t}) participating in a given stp. These stps are activated by outside signals P_{x+0} binding to membrane receptors (e.g., the glutamate receptors). Some of the proteins (P_{x+i}) enter the nucleus to control the gene reading, while other P_xs remain in the cytoplasm to control cellular events. Finally, other proteins (P_{x+t}) are exported to act on other cells or upon the environment.

From a formal point of view, a stp is a serial-ordered set of transactions:

$$p_{x+0} \to ... \to p_{x+i} \to .... \to p_{x+k} \to .... \to p_{x+t} \qquad (2.1)$$

where p_x stands for a given protein x, and $\to$ denotes some effect induced by p_x upon p_{x+1} due, for example, to some energy transference to, or structural modification of, p_{x+1}. For instance, the glutamate binding to its metabotropic receptor is proposed to expose an internal site of this receptor that splits a G-protein into its components G-α and G-γ, each one of which then controls a different chain of biochemical events. Because p_x act over different p_ys, which in turn may control different biochemical transactions, the order imposed upon any stp is a partial ordering. In some instances, the activation of a p_{x+1} may be dependent also on the action of proteins p_y, p_z besides p_x, that is:

$$p_y\, p_x\, p_z \to p_{x+1} \qquad (2.2)$$

2.1 Molecular Neurobiology

In this condition, it may be said that p_x acts upon p_{x+1} in the context (background) provided by p_y, p_z.

Biological remark 2.0. Note that an stp *is not an ordinary chemical reaction*, when two or more components react in order to generate one or more new chemical components. An stp is a sequence of energy transferences that convey some information to modulate the expression of a biochemical process supporting life. In such a way, it deserves a special kind of mathematical modeling, other than that classically applied to formal descriptions of the dynamics of chemical reactions.

The activity of any cell **C** at any moment is the result of a well orchestrated activation of many different stps. This may be described formally as:

$$C = \{P_o, P_n, P_t, B, D\} \quad (2.3)$$

Where:

1. P_o is the set of initial chemical signals p_{x+0};
2. P_n is the set of intermediate chemical signals p_{x+i};
3. P_t is the set of terminal chemical signals p_{x+t};
4. D is the set of genes encoding $P = P_o \cup P_n \cup P_t$, and
5. B is the set of biochemical transactions of the type $p_y\ p_x\ p_z\ p_y\ p_{x+i}, p_z$ denoting the interaction among the proteins $p_y\ p_x\ p_z$ which activates p_{x+i} in the context provided by p_y, p_z such that
6. the biochemical chain $stp(p_o, p_j)$ is characterized by the ordered set of transactions required to activate $p_j \in P_t$ whenever $p_o \in P_i$ is available, that is

$$stp(p_o, p_j) = p \rightarrow ... \rightarrow p_{x+i} \rightarrow ... \rightarrow p_{x+t} \quad (2.4)$$

Two different stp_i, stp_j are said to be a hierarchy if they share a subset of products P_c which are terminal products P_t of stp_i and initial products P_o of stp_j. In this condition, stp_i is said to regulate stp_j. Because early genes may coordinate the expression of different sets of downstream genes, then D becomes a hierarchical encoding H of $P^* = P_i \cup P_n \cup P_t$. H organizes families $\{stp_j\}_{j=1\ to\ n}$ of stps into a partially ordered set of biochemical chains (Fig. 2) supporting life. In the context of this paper, a Genetic Network (**GN**) is a set of these hierarchical stps supporting a given cellular activity:

$$GN = (\{stp_j\}_{j=1\ to\ n}, H) \quad (2.5a)$$

Where H may, for instance, be defined as:

$$H = \min \sum_{j-i}^{k} stp_j \quad (2.5b)$$

but many other kinds of hierarchies may also be proposed.

Biological remark 2.1. Central to the notions of stps and Genetic Networks (Fig. 2.2) is the concept of ambiguity in the molecular interactions in the chain of chemical reactions. Despite the specificity of many enzymes, many important proteins are involved in many stps, and in many instances the degree of molecular interaction is extremely variable, as in the case of transmitter and receptor binding, promoter binding to the TATA box, degree of protein phosphorylation, etc.

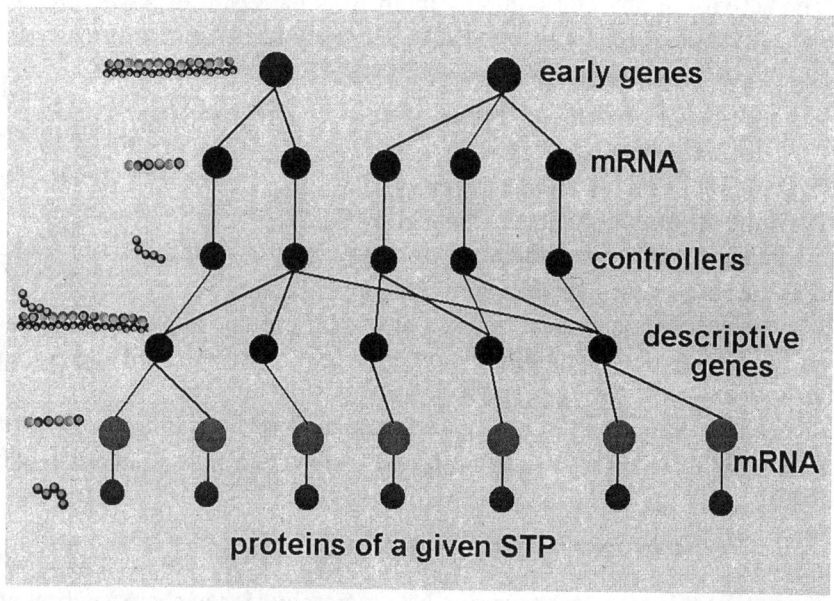

Fig. 2.2. A genetic network defining a stp: *a coherent set of genes is activated in order to produce all chemicals of a stp.*

One of the major challenges in the age of "Function Genomics" is to infer regulatory interactions between genes from experimental data collected from micro-array experiments. Genome expression analysis involves the use of oligonucleotide or cDNA microarrays to measure, in a parallel fashion, the mRNA levels of as many genes as possible in a genome. Many techniques are being developed to analyze these experimental measurements in order to disclose the main gene interactions in a given moment. Among these techniques, the Genetic Network has been used as a formalism to represent causalities among, and to reason about, these gene interactions. Again, many mathematical tools are being used to develop different kinds of genetic networks: boolean modeling (e.g., Akutsu et al. 2003); abductive reasoning (e.g., Zupan et al. 2003); multi-criterion optimization (e.g., van Smorerem et al. 2003), etc. Each of these techniques may be used to build H in Eq. 2.5.a.

From the present point of view, any GN is a structured set of information necessary to specify when and how one stp, or a group of stps, is to be expressed. A way of describing any GN is to provide a direct graph (Fig. 2.2), where the nodes represents elements of P* and the arcs describe the relations between pairs p_i, $p_j \in$ P*. Different types of functions may be associated with these arcs to describe the restrictions controlling the flow of information in the GN.

2.2 Fuzzy Formal Languages

Definition 2.1. A grammar **G** (e.g.; Chomsky, 1955; Mizumoto et al. 1973; Negoita and Ralescu, 1975; Rocha et al. 1980; Searls; 1992, 2002) is a structure defined as

$$G = \{V_o, V_n, V_t, P\} \quad (2.6)$$

where:

1. V_o is a set of initial or starting symbols;
2. V_t is a set of terminal symbols;
3. V_n is a set of non-terminal symbols, and
4. **P** is a set of rewriting rules defined as

$$p: \delta s_i \gamma \to \delta s_j \gamma, p \in P, \delta, \gamma, s_i, s_j \in V_o \cup V_n \cup V_t \cup \eta \quad (2.7)$$

In other words, p rewrites s_i as s_j in the context defined by δ and γ s_i, For the sake of simplicity, we denote

$$V^+ = V_o \cup V_n, V^\# = V_n \cup V_t, V^\& = V_o \cap V_t, \text{ and } V^* = V_o \cup V_n \cup V_t \quad (2.8)$$

V^* is supposed here to include the empty symbol ε.

The derivation chain $d(s_i, s_j)$ of the $s_i, s_j \in V^*$ is the ordered set of productions required to transform the symbol $s_i \in V_s$ into s_j. In other words

$$d(s_i, s_j) = \delta s_i \gamma \to \delta s_k \gamma \ldots \delta s_l \gamma \to \delta s_j \gamma \quad (2.9)$$

The grammar defined so far is called a type 0 grammar or G^0. Certain restrictions can be made on the nature of the productions of a grammar to give other types of grammars (Hopcroft and Ullman, 1969).

A language L(G) supported by G is the set of all n derivation chains $d(s_o, s_t)$, $s_o \in V_o$ and $s_t \in V_t$, that is

$$L(G) = \{d(s_o, s_t) = \delta s_o \gamma \rightarrow \delta s_i \gamma \ldots \rightarrow \delta s_t \gamma \mid s_o \in V_o \text{ and } s_t \in V_t\} \quad (2.10)$$

The processing of any derivation chain $d(s_o, s_t)$ is a partially sequentially ordered set of rewriting operations, each one involving the following steps:

1. **Matching**: a symbol at the left-hand side (e.g. s_k) of a prospective rewriting rule (e.g., $s_k \rightarrow s_j$) is matched ($s_i \equiv s_k$) to the symbols of the string s_i being processed. If this matching succeeds, then
2. **Rewriting**: the matched s_i is substituted by the right-hand side of the accepted rewriting rule $s_i \equiv s_k \rightarrow s_j$, and finally:
3. **Acceptance**: the membership degree $\mu V_t(s_i)$ of s_i to V_t is evaluated. If s_i is accepted as belonging to V_t, that is, if $s_i = s_t \in V_t$ the rewriting process is stopped, and $d(s_o, s_t)$ is assumed to be a well formed formula of L(G).

These steps describe the following:

2.1.a. Matching: a protein (symbol) s_k occurring on the left-hand side of a prospective biochemical reaction (or, rewriting rule) $s_k \rightarrow s_j$, is matched ($s_i \equiv s_k$) to the symbol (protein) s_i being processed (activated). The degree of matching $\mu(s_i \equiv s_k)$ is a measure of the chemical affinity (Tuszynski and Kurzynski 2003, pp 245-246) between s_i, s_k. In this way it may be defined as:

$$\mu(s_i \equiv s_k) = f\{q(s_i \equiv s_k) / [\min (q(s_i), q(s_K))]\} = f\{K * \exp(-\Lambda/\kappa T)\}$$

$$K = \exp(-\Delta\Omega^0 / RT)$$

that is, as a function f of Λ, the thermodynamic force or chemical affinity connected to the independent variable s_i, where κ is the Boltzmann constant; $\Delta\Omega^0$ is the free energy, T is the temperature in Kelvin degree; R is the gas constant; $q(s_i)$, $q(s_K)$, $q(s_i \equiv s_k)$ are the number of initial molecules (symbols) s_i, s_K and of $s_i \equiv s_k$, respectively. The function f is defined such that:

$$\mu(s_i \equiv s_k) \to 1 \text{ if } q(s_i \equiv s_k) \to \min(q(s_i), q(s_k)) \text{ and } \mu(s_i \equiv s_k) \to 0 \text{ if}$$
$$q(s_i \equiv s_k) \to 0$$

Now, if $\mu(s_i \equiv s_k) > 0$, then

2.1.b. Rewriting: the matched s_i triggers an action over s_k to activate s_j

$$s_i \equiv s_k \to s_j$$

and finally:

2.1.c. Acceptance: if s_j is either a molecule being exported outside the cell to act as a trigger of another stp in other cells, or an intermediary product of another stp in the same cell, then s_i is accepted as belonging to V_t; the rewriting process is stopped, and $d(s_o, s_t)$ is assumed to be a well formed formula of L(G).

Because a given s_k produced by a given $d(s_o, s_k)$ may be involved in more than one stp, it may be the case that it is an end product for one stp and an intermediate signal for another. For instance, ATP is a product in the case of respiratory stps, and intermediate components in many other biochemical processes, e.g., those illustrated in Fig. 2.

Along this line of reasoning, let the following be defined:

2.1.d: The **degree of similarity** $\varphi(s_i, s_k)$ of two strings s_i, s_k is a mapping in the Cartesian product space $V_s \times V_n \times V_t$ to the closed interval [0,1] such that:

$$\varphi : V_o \times V_n \times V_t \to [0, 1]$$
$$\varphi(s_i, s_k) = 0 \text{ if } s_i \neq s_k, \text{ otherwise } 0 < \varphi(s_i, s_k)$$

2.1.e: The **degree of acceptance** $\varphi(s_j, V_t)$ of s_k as belonging to V_t is calculated as the maximum degree of similarity $\varphi(s_j, s_t)$ of s_j with the strings $s_t \in V_t$. In other words:

$$\varphi(s_j, V_t) = \max \varphi(s_j, s_t) V_t$$

Now, let $s_k \in V^*$ be a chain which is defined over a set of characters given by $A = \{a_1 ..., a_n\}$, called a **fuzzy alphabet** (Sadegh-Zadeh, 2000) by means of a set of concatenation rules **F**:

$$F: (A \cup \eta)^l \to V^* \qquad (2.13)$$

such that each $s_k \in V^*$ is

$$s_k = \{ a_i \Leftrightarrow ... \Leftrightarrow a_k \Leftrightarrow ... \Leftrightarrow a_m \} \qquad (2.14)$$

or it is a concatenation ⇔ of characters of A with the maximum length equal to l, and

$$\mu_A: (A)^n \to [0, 1] \qquad (2.15)$$

measures the acceptability that a_l replaces a_k such that

$$\mu(a_k, a_l) = 1 \text{ if } s_k = \{a_i ... \Leftrightarrow a_k ... \Leftrightarrow a_m\} \text{ then } s_l = \{a_i ... \Leftrightarrow a_l ... \Leftrightarrow a_m\} \qquad (2.16)$$

In this context $\mu(a_k, a_l)$ is defined as a function μ_A that is

$$\mu_A: ((A)^n)^l \to [0, 1] \qquad (2.17)$$

In such a condition, A and F define the words s_k composing the sentences s_o accepted by L(G) if there exists $d(s_o, s_t)$ such that $\varphi(s_t, V_t) \to 1$.

Biological remark 2.2. The set of nucleotides composing the DNA and RNA, and the set of 20 amino acids forming the proteins, are examples of fuzzy alphabets. In this line of reasoning, $s_k \in V^*$ may be considered *words* composing the *stp sentences* of L(G) about the biochemical transactions supporting life. Genes are words encoding the proteins and proteins are also words used to compose the stps. In addition, genes are composed of two different substrings: the *code* substring containing the description of one or a family of proteins, and the *control* substring encoding when and how the gene is to be read. In such a way, Fuzzy Formal Language Theory is used here to deal with GN, in a context different from that used by Searls (1992, 2002). For instance, instead of focusing attention on the structure of the nucleotide sequences (words) in DNA, we discuss the dynamics of gene hierarchies and their role in specifying and organizing the (sentences) stps. Most of what follows deals with the stp *sentence* dynamics, rather than the structure of both the genetic and proteomic *words*.

2.3 Ambiguity

Fuzzy languages exhibit distinctive properties compared to crisp languages because, given $s_i \in V^*$, many $s_k \in V^*$ may exist for which $\mu(s_i \equiv s_k) > 0$. This means that many derivation chains $d(s_o, s_k)$ may rewrite s_o into many different $s_t \in V_t$, depending on the ambiguity of any among its rewriting steps $s_i \equiv s_k \to s_j$. The ambiguity of L(G) depends on how many derivation strings $d(s_o, s_t)$ exist for the same $s_k \in V_o$ resulting in different $s_j \in V_t$. Therefore, Fuzzy Grammars are naturally ambiguous and the amount of ambiguity, $\Omega(G)$, of a given grammar G is strongly related to the cardinalities of V_o, V_i, V_t and to the distribution of the actual values of $\mu(s_i, s_k)$ in its state space $V_s \times V_n \times V_t$.

The usefulness of a fuzzy grammar G is dependent on how $\Omega(G)$ is constrained by the resources available to process G. These constraints may be imposed upon, for instance, the dictionary A generating the words s_k of L(G), in such a way that $q(s_k)$ becomes dependent on the availability $q(a_i)$ of $a_i \in s_k$. It may also be dependent on any control imposed upon F in Eq. 2 13. In this context of, the ambiguity $\Omega(G)$ of G is assumed to be dependent on:

1. the total number or quantity, $q(s_i)$, of available copies of $s_i \in V^*$: at least one copy of $s_i \in V^*$ has to be available to trigger each possible derivation chain $d(s_i, s_j | s_k)$, supported by $\delta s_k \gamma \to \delta s_j \gamma$, $\mu(s_i \equiv s_k) > 0$,
2. the total number, $q(s_k)$ of copies of $s_k \in V^*$: at least one copy of $\delta s_k \gamma$ has to be available to allow the rule $\delta s_k \gamma \to \delta s_j \gamma$, $\mu(s_i \equiv s_k) > 0$ to be used, and
3. the total number, $q(s_j)$ of copies of $s_j \in V^*$: at least one copy of s_j has to be available to allow s_i to be rewritten into s_j.

In this context:

Definition 2.2.a: The **possibility** $\rho(d(s_i, s_j) | H)$ that the derivation chain $d(s_i, s_j | s_k)$ supported by $s_i \equiv s_k \to s_j$ is used to rewrite s_i into s_j, under the constraints imposed by the environment H is a function of $\mu(s_i \equiv s_k)$ and the chemical affinity associated to s_j (Tuszynski and Kurzynski, 2003, pp 249-250). In this way, it is calculated as:

$$\rho(d(s_i, s_j | s_k) | H) = \mu(s_i \equiv s_k) \otimes [q(s_j) / q(s_i)^* q(s_k)]$$

$$[q(s_j) / q(s_i)^* q(s_k)] = K^* \exp(-\Lambda / \kappa T)$$

$$K = \exp(-\Delta\Omega^0 / RT) \tag{2.18}$$

where n is the cardinality of V_n, and $\otimes$ is a T-norm (see Pedrycz and Gomide, 1998, about the concept of S and T-norms. The max and min functions are examples of S and T-norms, respectively).

In such conditions, $\rho(d(s_i, s_j | s_k) | H)$ may assume any value in the closed interval [0, 1].

For the sake of simplicity, $\mu(s_i \equiv s_k)$ is denoted as $\mu(s_i, s_k)$ in the rest of the paper.

Definition 2.2.b: The possibility $\rho_k(d(s_o, s_t))$ of a given
$$d(s_o, s_t | s_k) = \delta s_0 \gamma \ldots \to \delta s_k \gamma \to \delta s_1 \gamma \ldots \to \delta s_t \gamma$$
including a defined $\delta s_k \gamma \to \delta s_1 \gamma$, is determined by the weakest rewriting step. In other words:

$$\rho(d(s_i, s_j | s_k) | H) = \Psi_{k=1}^{n} \{\mu(s_i \equiv s_k) \otimes [q(s_j)/q(s_i) * q(s_k)]\} \tag{2.19a}$$

where Ψ is an S-norm, it will be assumed as the min operator in the rest of this chapter.

Definition 2.2.c: The possibility:
$$\rho(d(s_o, s_t)) \text{ of } d(s_o, s_t) = \delta s_o \gamma \ldots \to \delta s_t \gamma$$
is
$$\rho(d(s_o, s_t)) = \max \rho(d(s_o, s_t) | k) \tag{2.19b}$$

because ambiguity in G creates many alternative pathways $d(s_o, s_t | s_k)$ whereby one can derive s_t from s_o. A given $d(s_o, s_t)$ is said to be expressible if $\rho(d(s_o, s_t)) > 0$.

Proposition 2.1. The constraints imposed by H define a unique distribution $\Pi(\rho(d(s_o, s_t) | H))$ over V^* such that the mean possibility $<\rho(d(s_o, s_t))>$

$$<\rho(d(s_o, s_t))> = 1/m \sum_{j=1}^{m} \rho(d(s_o, s_t)) \tag{2.20}$$

is subject to
$$<\rho(d(s_o, s_t))> \to \lambda, \ 0 < \lambda < 1 \tag{2.21}$$

where m is the cardinality of V^*.

Proof: It follows from Definition 1, in order to guarantee the existence of at least one $d(s_o, s_t | s_k)$ such that $\rho(d(s_o, s_t | s_k) | H) > \lambda$; thus:
$$\rho(d(s_o, s_t | s_k) | H) > 0.$$

Biological remark 2.3. H is assumed here to stand for the set of Earth's conditions supporting the existence of a set of living beings defined by **G**. The restrictions imposed by H may be upon the availability of:

1. **the elements of the alphabet A**: that is, upon the availability of nucleotides, amino acids, etc;

2. **the energy to support the chemical STP transactions**: that is, upon the energy required for protein phosphorylation;
3. **other chemicals, such as ions, required to support STP transactions**: that is, upon the availability of ions such as Ca^{2+} used as a second messenger; or Na^+, K^+ and Cl^-, specifying the electrical cell environment or its pH, etc.

Life is characterized by chemical changes in an inevitable progression, which is not due to blind chance. The chemical sequence is from a reducing, to an ever increasing, oxidizing environment wherein organisms struggle to retain reduced chemicals. Most of this dynamics was determined by the environmental conditions of primeval earth, when biodiversity began to blow up (Williams and Fraústo da Silva, 2003). The restrictions imposed by H are one of the driving forces of evolution, the other one being the incompleteness or the genetic information specifying any living being (Barbieri, 2003). Environmental randomness and $d(s_i, s_j)$ ambiguity will conjoin forces to drives life evolution.

Since the focus of the present chapter is on the informational contents of the biochemical transactions supported by the stp sentences, given a certain $d(s_o, s_t)$ supported by the grammar G in H, and $s_i \equiv s_k \rightarrow s_j$, let the following to be defined:

1. **entropy**: $h(d(s_o, s_t | s_k) | H)$ as

$$h(d(s_o, s_t) | H) = \qquad (2.22)$$
$$- (\rho(d(s_o, s_t)) \log \rho(d(s_o, s_t)) - (1 - \rho(d(s_o, s_t))) \log (1 - \rho(d(s_o, s_t)))$$

2. **mean ambiguity**: $<\Omega d(s_o, s_t) | H>$ as

$$<\Omega(d(s_o, s_t) | H)> = - (<\rho(d(s_o, s_t))> \log <\rho(d(s_o, s_t))> \qquad (2.23)$$
$$- (1 - <\rho(d(s_o, s_t))>) \log(1 - <d(s_o, s_t))>))$$

3. **expressiveness**: $\theta(d(s_o, s_t | s_k) | H)$ of $d(s_o, s_t | s_k)$ as

$$\theta (d(s_o, s_t | s_k)) = <\Omega (d(s_o, s_t))> - h(d(s_o, s_t | s_k) | H) \qquad (2.24)$$

In this context, the expressiveness of $L(G | H)$ is calculated as

$$\theta(L(G | H)) = \sum_{i=1}^{n_{G|H}} \theta(d_i(s_o, s_t | s_k)) \qquad (2.25)$$

where $n_{G|H}$ is the number of $d_i(s_o, s_t | s_k)$ for which $\rho(d(s_o, s_t | s_k) | H) > 0$.

Note that entropy here is a measure of both the matching ($s_i \equiv s_k$) uncertainty and the uncertainty in word frequency, because from Eq. 2.18 $\rho(d(s_i, s_j))$ is a function of both $\mu(s_i, s_k)$ and $q(s_i), q(s_k), q(s_j)$.

Theorem 2.1. $\theta(L(G \mid H)) \rightarrow n_{G|H}$ bits if $\lambda \approx 0.5$ in Eq. 2.21.

Proof: If $<\rho(d(s_o, s_t))> = 0.5$ then

$$\theta(d(s_o, s_t \mid s_k)) = 1 - h(d(s_o, s_t \mid s_k) \mid H)$$

and

$$\theta(L(G|H)) = n_{G|H} - \sum_{k=1}^{n_{G|H}} h_i(d(s_o, s_t \mid s_k) \mid H)$$

Since

$$h(d(s_o, s_t \mid s_k) \mid H) \rightarrow 0 \text{ if } \rho(d_i(s_o, s_t \mid s_k)) \rightarrow 1 \text{ or } 0$$

then

$$\Sigma\, h_i(d(s_o, s_t \mid s_k) \mid H) < n_{G|H}$$

Therefore

$$\theta(L(G \mid H)) \rightarrow n_{G|H} \text{ bits.}$$

as the number $d_i(s_o, s_t \mid s_k) \mid \rho(d_i(s_o, s_t \mid s_k)) \rightarrow 1$ or 0 increases, while maintaining $\lambda \approx 0.5$.

This is accomplished if for each $\rho(d_i(s_o, s_t \mid s_k)) \rightarrow 1$ there is another $\rho(d_i(s_o, s_t \mid s_k')) \rightarrow 0$ such that

$$\rho(d_i(s_o, s_t \mid s_k)) + \rho(d_i(s_o, s_t \mid s_k')) \rightarrow 1$$

Corollary 1: $\theta(L(G \mid H)) \rightarrow \xi$ bits, $\xi \leq 0$ as $\lambda \rightarrow 1$ or 0 in Eq. 2.21.

Proof: If $<\rho(d_i(s_o, s_t))> \rightarrow 1$ or 0 then

$$\theta(d(s_o, s_t \mid s_k)) = \tau - h(d(s_o, s_t \mid s_k) \mid H), \tau \rightarrow 0$$

and

$$\theta(L(G \mid H)) = \tau^* n_{G|H} - \sum_{k=1}^{n_{G|H}} h_i(d(s_o, s_t \mid s_k) \mid H)$$

Since

$$h(d(s_o, s_t \mid s_k) \mid H) \rightarrow 1 \text{ if } \rho(d_i(s_o, s_t \mid s_k)) \rightarrow 0.5,$$

otherwise

$$h(d(s_o, s_t \mid s_k) \mid H) \rightarrow 0.$$

Then

$$\sum_{k=1}^{nG|H} h_i(d(s_o, s_t \mid s_k) \mid H) \to \sigma \geq 0$$

and

$$\theta(L(G \mid H)) = \tau^* \, n_{G|H} - \sigma$$

such that

$$\theta(L(G \mid H)) \to \xi^* \text{ bits}, \ \xi \leq 0$$

Biological remark 2.4. The expressiveness $\theta(L(G \mid H))$ of a given language $L(G \mid H)$ supported by G in H is a key issue in understanding adaptation and evolution. Evolution and adaptation are allowed in those cases where $L(G \mid H)$ is expressible. In contrast, extinction will imply the expressiveness is greatly reduced. The actual value of $\theta(L(G \mid H))$ is mostly determined by the restrictions imposed by H, and determining $\Pi(\rho(d(s_i, s_j) \mid H))$. The restrictions imposed by H are among the driving forces of evolution, because they control what the expressible $d(s_o, s_t \mid s_k)$ are that shape a living being. The incompleteness of the genetic code defines the possible expressible $d(s_o, s_t \mid s_k)$s. Because of this, $\Pi(\rho(d(s_i, s_j) \mid H))$ becomes a general formalization of *H* in Eq. 2.5.a.

Definition 2.3: The space S in the environment H is said to be a processing space for the fuzzy grammar G if it guarantees $\theta(L(G \mid H)) > 0$ that is

$$V_t(H \mid S) = \{s_t \in V_t \mid (\rho(d(s_o, s_t) \mid H, S) \to 1)\} \neq \phi$$

If $\Pi(\rho(d(s_k, s_j) \mid H))$ is inhomogeneous in S, the processing space S can be partitioned into a set of n subspaces $\{S_m\}_{m=1 \text{ to } n}$ that become specialized spaces for processing defined sets of derivations chains $d(s_k, s_j)$ of G. Therefore, each S_m may become a specialized space for processing defined $\rho(d(s_o, s_t \mid s_k))$.

Biological remark 2.5: The concept of processing space S is not to be mistaken with the geographic location 1 of H, but it must be assumed as a defined physical subspace, corresponding, for example, to that occupied by an organism's body. In this context, S_m, S_n will denote two distinct elemental bodies, or cells.

Definition 2.3.a: Two processing spaces S_m, S_n for the fuzzy grammar G are said to be compatible $S_m \equiv S_n$ processing spaces of G if and only if $\Pi(\rho(d(s_i, s_j) \mid H, S_m)), \Pi(\rho(d(s_i, s_j) \mid H, S_n))$ are such that

$$V_t(H \mid S_m) \cap V_t(H \mid S_n) \neq \phi.$$

Definition 2.3.b: The degree of compatibility $\mu(S_m, S_n)$ of S_m, S_n as processing spaces of G in H is measured by the similarity $\mu(V_t(H \mid S_m), V_t(H \mid S_n))$ between $V_t(H \mid S_m)$ and $V_t(H \mid S_n)$ such that

$$\mu(S_m, S_n) = \mu(V_t(H \mid S_m), V_t((H \mid S_n)) = c_{i,j} \div \max(c_i, c_j)$$

where $c_{i,j}$ is the cardinality of $V_t(H \mid S_m) \cap V_t(H \mid S_n)$, and c_i, c_j are the cardinality of $V_t(H \mid S_m)$ and $V_t(H \mid S_n)$, respectively.

Definition 2.3.c: If the transcription $\delta s_i \gamma \to \delta s_j \gamma$ is performed at the time t_i at the processing space S_m of G, the result of this transcription will be available at the time $t_j = t_i + \Delta$ at the processing space S_n of G, where Δ is the finite time restriction imposed by H. Also, let

$$<\rho(d(s_i, s_j) \mid H, S_m, S_n, \Delta)>$$

be the mean value of

$$\rho(d(s_i, s_j) \mid H, S_m, S_n, \Delta(t))$$

of moving all $s_i \in V^*$ from S_m to S_n in H.

Biological remark 2.6. The temporal restrictions Δ imposed by H are mainly those associated with the dynamics of the intra and intercellular transportation systems in charge of moving proteins inside and between cells, and DNA and RNA between cell compartments.

2.4 The Hierarchy of Fuzzy Grammars

According to Searls (2002), the dynamics of both genes and proteins must be modeled by grammar above the level of the context sensitive (Type 1 grammar - Hopcroft and Ullman, 1969) grammars in the Chomskyan hierarchy. This is one of the reasons for introducing the following hierarchy of the Recursive Enumerable Grammars:

1. Replicating Grammars: to formalize properties and consequences of DNA duplication;
2. Self-controlled Grammars: to provide the tools to control grammar ambiguity and to improve adaptability; and
3. Recombinant Grammars: to formalize properties and consequences of the sexual reproduction to life evolution.

Definition 2.4. Replicating Grammars $G^{\circledR}$: Let there be the grammar

$$G^{\circledR} = \{V_0, D \subset V_n, V_t, R \subset P, \eta) \tag{2.26}$$

where

2.4 The Hierarchy of Fuzzy Grammars

$$D = \{s_d \in D \mid \rho(d(s_d, s_m)) > 0 \text{ for each } s_y \in V^*\} \quad (2.27)$$

and

$$R = \{\delta s_d \gamma \rightarrow \delta s_1 \gamma \ldots \delta s_i \gamma \rightarrow \delta s_d \gamma, s_d \in D\} \quad (2.28)$$

D is said to encode the symbols of G under the rules R. Both D and R defines de genetic G describing V*, that is

$$G^® = (D, R, V^*) \quad (2.29)$$

The genetic G allows $G^®$ to be copied or duplicated from the processing space S_m to S_n in H if for all $s_d \in D$ such that $\rho(d(s_d, s_d) \mid H_m) \rightarrow 1$ there exists Δ such that $\rho(d(s_d, s_d) \mid H, S_m, S_n, \Delta) \rightarrow 1$.

Biological remark 2.7.a. D is the set of genes encoding the genetic information of a given living being. Any gene s_d is composed by two substrings, one of the encoding one or more proteins s_k of this living being, and the other being part of the gene reading control mechanism. In this context, $\Pi(\rho(d(s_d, s_d) \mid H, S_m))$ describes the constraints over the process of gene copying and $\Pi(\rho(d(s_d, s_d) \mid H, S_m, S_n, \Delta))$ describes the constraints over the mitosis and meiosis at the cellular level. In the same line of reasoning, $\Pi(\rho(d(s_d, s_k) \mid H, S_m))$ depends on the constraints imposed upon the gene reading in the process of protein making, whereas the constraints on protein making are encoded by $\Pi(\rho(d(s_d, s_d) \mid H, S_m, S_m, \Delta))$ in the same processing space (cell) S_m.

Let $D(H \mid S_i)$, $D(H \mid S_m)$ be the original and copied fuzzy sets of s_d s of $G^®$. Since the ambiguity of $G^®$, may produce $s_d' \in D(H \mid S_m)$ as a copy of $s_d \in D(H \mid S_m)$, whenever $\mu(s_d', s_d) > .5$, the fidelity $\mu_G^®(H \mid S_m, S_n)$ of the duplication of $G^®$ from S_m into S_n in H is defined here as

$$\mu_G^®(H \mid S_m, S_n) = \max_{\alpha=0.5}^{1} (\alpha * d \div \max(c_i, c_j)) \quad (2.30)$$

where **d** is the cardinality of $D_\alpha(H \mid S_m) \cap D_\alpha(H \mid S_n)$, and c_i, c_j are, respectively, the cardinalities $D_\alpha(H \mid S_m)$ and $D_\alpha(H \mid S_n)$. The actual values of d, c_i, c_j are dependent on both

$$\Pi(\rho(d(s_d, s_d) \mid H, S_m)) \text{ and } \Pi(\rho(d(s_d, s_d) \mid H, S_m, S_n, \Delta))$$

because

$$\mu_G^®(H \mid S_m, S_n) \rightarrow 1$$

iff $\rho(d(s_d, s_d) \mid H, S_m, S_n, \Delta) * \rho(d(s_d, s_d) \mid H, S_m) \rightarrow 1$

2 The Cells of the Brain

Thus it may be assumed that

$$\mu_G^{\circledR}(H \mid S_m, S_n) = \rho(d(s_d, s_d) \mid H, S_m, S_n, \Delta) * \rho(d(s_d, s_d) \mid H, S_m) \quad (2.30b)$$

Biological remark 2.7.b. Gene mutation is defined when a gene copy s_d' of s_d is such that $\varphi(s_d', s_d) \to 0$. The set D of $G^{\circledR}$ corresponds to the genome of any living being, and the notion of gene mutation in this paper is a consequence of the ambiguity of $G^{\circledR}$, in addition to possible errors in gene copying. Therefore, gene mutation is dependent on the behavior of $\rho(d(s_d, s_d) \mid H, S_m)$. In this context, the rate of mutations is inversely related to $\mu_G^{\circledR}(S_i, S_j \mid H)$, because d in Eq. 2.30 will decrease as the number of mutation augments. Also, the genetic information (D, S_n) will be modified in respect to (D, S_m) if $\rho(d(s_d, s_d) \mid H, S_m, S_n, \Delta)$ is changed. In this context, the rate of possible alterations of (D, S_m) is inversely related to $\mu_G^{\circledR}(S_i, S_j \mid H)$.

Definition 2.5. Self-controlled Grammar: Let there be a replicating grammar

$$G^{\circledcirc} = \{V_o, V_n, C \subset V_t, E \subset P, \eta\} \quad (2.32)$$

C is the set of $s_t \in V_t$ that controls the amount of copies $q(s_i)$; $q(s_j)$ or the matching capability $\mu(s_i, s_j)$ of $s_i, s_j \in V^*$, that is

$$\{s_c \in V_t \mid [(q(s_i); q(s_j) = g(q(s_c)), \mu(s_i, s_j) = z(q(s_c)), s_i, s_j \in V^*]\} \quad (2.33)$$

A given $s_c \in C$ exerts its control over the $q(s_j)$ of $s_j \in V^*$ by:

1. promoting the decoding of s_d: in this case

$$d(s_d, s_j) = \delta s_d \gamma \to \delta s_c \gamma \to \ldots \to \delta s_j \gamma, \text{ or}$$

2. enhancing or inhibiting the decoding of s_d: in this case s_C substitute more (or less) efficiently another $d(s_k, s_l)$ of $d(s_d, s_j)$:

$$d(s_d, s_j) = \delta s_d \gamma \to \ldots \to \delta sk_d \gamma \to \delta s_l \gamma \to \ldots \to \delta s_j \gamma$$
$$d(s_d, s_j) = \delta s_d \gamma \to \ldots \to \delta s_c \gamma \to \delta s_m \gamma \to \ldots \to \delta s_j \gamma$$

3. changing $\mu(s_i, s_k)$ of any intermediate $\ldots \to \delta s_i \gamma \to \delta s_j \gamma \to \ldots d(s_d, s_j)$: in this case s_c becomes an intermediate step of $d(s_i, s_k)$ of $d(s_d, s_j)$

$$d(s_i, s_j) = \delta s_i \gamma \to \delta s_c \gamma \to \delta s_j \gamma$$

As a matter of fact, the above actions may be exerted over any $d(s_o, s_j)$ supported by $G^{\circledcirc}$.

Biological remark 2.8. In the case of genes, the actions in a and b are those ascribed to the promoters, enhancers and inhibitors of the DNA reading, whereas that of action d is that exerted by polymerase A. In the case of any other stp, the actions a and b correspond to the triggers, facilitators and inhibitors of any of its steps, whereas the action d corresponds to that of those enzymes phosphorylating the proteins in this stp. The control allowed by C over $\rho(d(s_i, s_j))$ is the main tool Nature has invented in order to enhance adaptation over an increasing number of different environments Hs. This is because C allows the organism S_k to try to maintain $\Pi(\rho(d(s_i, s_j)|\ H, S_k))$ as constant as possible, where the restrictions imposed by the different H's would make survival unlikely.

Definition 5.a. For each

$$d(s_o, s_j \mid s_k) = \delta s_o \gamma \rightarrow ... \rightarrow \delta s_i \gamma \rightarrow \delta s_k \gamma \rightarrow ... \rightarrow \delta s_j \gamma \qquad (2.34)$$

there exists a set of genes describing their symbols

$$D = \{s_d \in D \mid d(s_d, s_k) \text{ for each } s_k \text{ in Eq. 34}\} \qquad (2.35)$$

the control set C of D

$$C = \{s_c \in C \mid (q(s_d) = g(q(s_c))\} \qquad (2.36)$$

and the set of genes specifying C

$$E = \{s_e \in D \mid d(s_e, s_c) \text{ for each } s_c \in C\} \qquad (2.37)$$

The set E is called here the set of early genes, each of them controlling the expression of a given $c \in C$, which in turn controls the expression of the descriptive genes in D in charge of specifying the other elements $s_k \in V^*$. Eqs. 34 to 37 comprise a hierarchical fuzzy tree whose purpose is to guarantee the existence of all s_k involved in a given $d(s_o, s_j)$.

Proposition 2.2: The set of early genes E specifies the root node set of a fuzzy tree GN_i and are connected to those first order of intermediary nodes are associated to C, which in turn are linked to the set of nodes representing D. The D nodes are finally linked to the terminal nodes representing each s_k involved in $d(s_o, s_t)$. GN_i is called here the genetic network controlling $d(s_o, s_t)$. N_i specifies the grammar of such control, which is considered to be a sub-grammar of $G^©$.

Proof: This is a consequence from the fact that E is in charge of controlling of the expression of D and thus of the dynamics of a serial ordered chain of rewriting rules $\delta s_i \gamma \rightarrow \delta s_k \gamma$. Therefore, E imposes a hierarchy H over $d(s_o, s_t)$.

In this context, the genetic $G^©$ supported by D, R may be considered as a family of GN_i each one controlling one of the m families $\{d(s_o, s_t)\}_{k=1 \text{ to } n}$ of stps supported by $G^©$ and specifying a given cellular function f_m:

$$G^{\copyright} = \{GN_i\}_{i=1 \text{ to } m} \quad (2.38)$$

The resources constraints imposed by H over $G^{\copyright}$ implies

$$q(s_d) \to 1 \quad \text{for each } s_d \in GN_i \quad (2.39)$$

and the number of copies $q(GN_i)$ of GN_i to be maintained as minimum as possible

$$1 \leq q(N_i) < \theta \quad \text{for each } GN_i \in G^{\copyright}, \theta \to 2 \quad (2.40)$$

Biological remark 2.9. All living beings (some viruses being perhaps the only exceptions), are examples of the self-controlled grammar $G^{\copyright}$ generated from the replicating grammar $G^{\circledR}$ supported by DNA. Viruses are considered an exception because they use the control machinery of the infected cell. In this paper, self-control will be proved to be a very important tool to both increase adaptation and influencing the evolution of the species. The complexity of $G^{\copyright}$ greatly increases in respect to that of $G^{\copyright}$ because the structure of each $GN_i \in G^{\copyright}$ describes a sub-grammar specifying how and when any stp is produced and organized as a specific combination of proteins belonging to V_t.

Definition 2.6. Recombinant Grammars: Let there be the self-controlled grammar

$$G^{\circledR} = \{V_o, D^{\circledR} \subset V_n, V_t, R, M \subset P, \eta\} \quad (2.41)$$

where $M \subset P$ is a set of rules allowing:

$D^{\circledR}$ to be divided into new $D^{\female}, D^{\male}$, such that $q(N_i) \to 1$ at $D^{\female}, D^{\male}$ and two copies:

$$G^{\female}(S_m) = \{V_o, D^{\female} \subset V_n, V_t, R, M \subset P, \eta \mid q(N_i) \to 1 \; \forall N_i \in G^{\circledR}\} \quad (2.42)$$

$$G^{\male}(S_m) = \{V_o, D^{\male} \subset V_n, V_t, R, M \subset P, \eta \mid q(N_i) \to 1 \; \forall N_i \in G^{\circledR}\}$$

of $G^{\circledR}(S_m)$ at S_m are move to two different sub-processing spaces of $S_m^{\female}$, $S_m^{\male}$ in H, and the recombination $G^{\female}(S_m) \leftarrow G^{\male}(S_n)$ of $G^{\female}(S_m), G^{\male}(S_n)$ from two different sub-processing spaces $S_m^{\female}, S_n^{\male}$ into $G^{\circledR}(S_s)$ such that

$$D_s^{\circledR} = D_m^{\female} \cup D_n^{\male}$$

and

$$G^{\circledR} = \{N_i^{\circledR}\}_{i=1 \text{ to } c'} \cup \{N_i^{\circledR}\}_{i=1 \text{ to } c''}, c', c'' \to c \text{ the cardinality of } G_i^{\circledR}$$

Biological remark 2.10. The recombinant grammars $G_i^{\circledR}$ generated from the DNA replicating grammar $G^{\circledR}$ support sexual reproduction and greatly facilitates the increase of the complexity of $G^{®}$ required to promote complex animals, such as man, in evolution. The steps in 1 are the main steps of meiosis, and those in 2 are the main events of fecundation. $G^{♀}(S_m)$ and $G^{♂}(S_m)$ are the female and male gametes, respectively.

For the sake of simplicity, G will denote in the rest of this article any grammar $G^{®}$, $G^{©}$ or $G^{®}$, such that any of these special notations will be used only if necessary to stress properties not shared by all these grammar types.

2.5 Fuzzy Languages, Distributed Processing and Biological Diversity

4.1. Definition: A *language* L(G) supported by G is composed by the set of all $d(s_o, s_t)$ converting $s_o \in V_o$ into $s_t \in V_t$ and can be defined as

$$L(G) = \{d(s_o, s_t) \mid s_o \in V_o, s_t \in V_t\} \tag{2.43}$$

Any $d(s_o, s_t), d(s_o, s_t')$ are said to be equivalent interpretations of s_o in L(G), that is $d(s_o, s_t) \equiv d(s_o, s_t')$, if $\varphi(s_t, s_t') \to 1$

Biological remark 2.11. Each $d(s_o, s_t \mid H, S)$ is supposed to describe a given stp in the real cell.

Definition 2.7. The hypothetical L(G) is said to be an expressible language $Ł(G \mid H, S)$ in the space S if and only if there exists at least one H providing $\Pi(\rho(d(s_o, s_t) \mid H, S))$ such that $\rho(d(s_o, s_t)) \to 1$ for at least one $s_o \in V_o$. In this context

$$Ł(G \mid H, S) = \{d(s_o, s_t) \mid \rho(d(s_o, s_t) \mid H, S) \to 1)$$

$$V_o(L \mid H, S) = \{s_o \in V_o \mid \rho(d(s_o, s_t) \mid H, S) \to 1 \text{ for } \in V_t\}$$

$$V_t(L \mid H, S) = \{s_t \in V_t \mid \rho(d(s_o, s_t) \mid H, S) \to 1 \text{ for all } s_o \in V_o(L \mid H, S)\}$$

$$V_n(L \mid H, S) = \{s_i \in V_n \mid \delta s_i \gamma \text{ is part of at least one } d(s_o, s_t) \text{ of } Ł(G \mid H, S)\}$$

$V_o(L \mid H, S)$ is the set of all expressible $s_o \in V_o$ and $V_t(L \mid H, S)$ is the set of expression of L(G) at S, given H. $Ł(G \mid H, S)$ is composed of all expressible derivations chains $d(s_o, s_t)$ for which $\rho(d(s_o, s_t) \mid H, S) \to 1$.

If $S_m, S_n \in S$ are processing subspaces $Ł(G \mid H, S)$, such that

$$V_t(L \mid S_m) \cap V_o(L \mid S_n) \neq \phi \ , \ V_o(L \mid S_m) \cap V_t(L \mid S_n) \neq \phi \tag{2.44}$$

then the processing of $\mathcal{L}(G \mid H, S)$ is said to be distributed over $\{S_m, S_n\}$ or $\mathcal{L}(G \mid H)$ is said to be a distributed processing language.

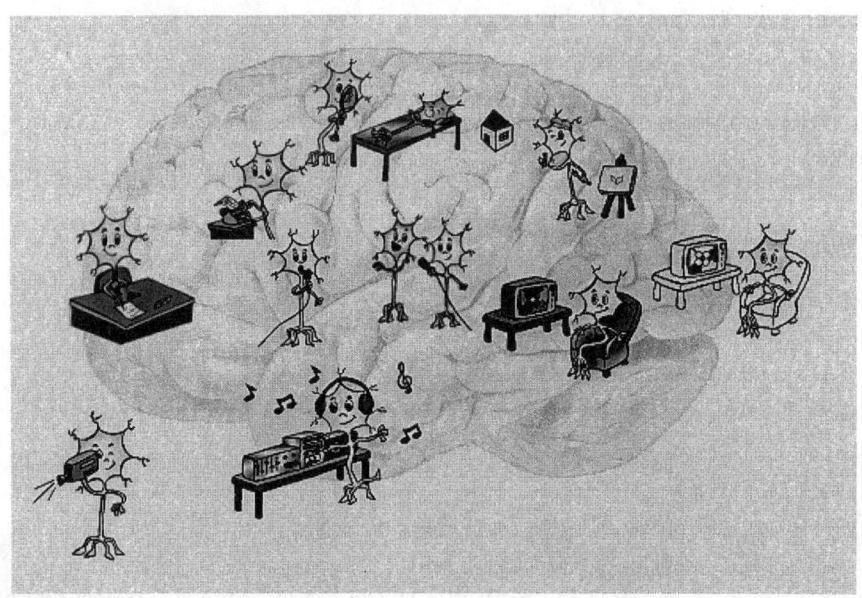

Fig. 2.3. The brain as a distributed G processor: *different types of neurons are specialized to process different subsets of L(G).*

Remark 2.12.a. The several expressed languages $\mathcal{L}(G \mid H, S)$ of the grammar G, supported by the DNA replicating grammar G, correspond to the different types of living beings existing or having existed on earth. In this context, $\mathcal{L}(G \mid H, S_i)$ the language characterizing a type of living entity S_i. In the case of unicellular organisms S^u or colonies of S^us, the processing space S_u is unique. Now, if $S = \{S^m{}_k\}_{k=1 \text{ to } n}$ then $\mathcal{L}(G \mid H, \{S^m{}_k\}_{k=1 \text{ to } n})$ is a multi-cellular organism S^m, composed of n types of cells and expressing n types of languages $\mathcal{L}(G \mid H, \{S_k\})$ of those supported by G. These different types of cells are part of the different organs in a multi-organ being. Each organ will contain many cells of each type; thus, each organ express a family of languages $\mathcal{L}(G \mid H, \{\{S_{k,m}\}_{k=1 \text{ to } n}\}_{m=1 \text{ to } r})$, where n is the number of types of its cells and m is the number of each of these cell types. The brain B is one of these organs and it is constituted of a subset of these processing spaces $S_{i,k}$; that is, $B = \{\{S_{i,k}\}_{i=1 \text{ to } n}\}_{k=1 \text{ to } m}$ is composed of m neurons of n types. In this context $\mathcal{L}(G \mid H, B)$ is the knowledge an animal may use to survive in H.

The constraint $\rho(d(s_o, s_t) | H, S) \to 1$ implies that $\mu(s_t, V_t)$ is bounded into a limited interval or

$$\mu(s_t, V_t) \geq \alpha_t \ \forall \ s_t \in V_t(L | S), \text{ where } \alpha_t = \xi_t \ \& \ 0 < \xi_t < 0.5 \quad (2.45)$$

This means that $V_t(L | H, S)$ is the α_t cut level set of V_t (Pedrycz and Gomide, 1998) having at least one $s_o \in V_o$ and one $s_t \in V_t$ such that $\rho(d(s_i, s_t) | H, S) = 1$. Because of this, each $L(G | H, S)$ is uniquely identified by its characteristic set

$$V_t^@ (L | H, S) = \max_{\alpha=\beta}^{1} \{s_t \in V_t | \rho(d(s_o, s_t)) | H, S) = \alpha_t\} \beta \to 1, \quad (2.46)$$

the maximum non-empty α_m cut level set of V_t. The power of $L(G | H, S)$ is then the number p of those $d_j(s_o, s_t)$, $\rho(d(s_o, s_t)) \geq \alpha_m \to 1$ which support $V_t(L | H, S)^@$.

The similarity of two languages $\mu(L(G | H, S), L(G | H', S'))$ is calculated as

$$\mu(L(G | H, S), L(G | H', S')) = l_{i,j} \div \max (l_i, l_j) \quad (2.47)$$

where $l_{i,j}$ is the cardinality of $V_t^@ (L | H, S) \cap V_t^@ (L | H', S')$ and l_i, l_j are the cardinalities of $V_t^@ (L | H, S)$ and $V_t^@ (L | H', S')$, respectively.

Biological remark 2.12.b. Any $L(G | H, S)$ is supposed to support a family of similar, but not identical, living beings $L(G H, S^i)$ as a consequence of the ambiguity in G. This diversity is a very important issue in understanding evolution and speciation.

2.6 Knowledge Adaptation and Evolution

The ambiguity of G, together with the restrictions imposed by H over $\Pi(\rho(d(s_i s_j) | H))$ will be used here to model the evolution of the expressed languages $L(G | H, S)$ supported by the grammar G in the attempt to understand how learning is accomplished by the brain B as a distributed processor of G.

Theorem 2. 2. Given $L(G | H, S)$ as a expressed language of G in H and $L(G | H', S')$ as a possible expression of G in H' if $| \lambda_{H'} - 0.5 | \leq | \lambda_H - 0.5 |$ then $L(G | H', S')$ is adapted to H'.

Proof: From corollary 2.1, $\theta(L(G|\,H',S'))$ decreases and $L(G|\,H',S')$ becomes inexpressible as $\lambda_H \to 0$ or 1, and theorem 2.1 shows that $\theta(L(G\,|\,H',S'))$ increases as $\lambda_H \to 0.5$.

Hence if $|\lambda_{H'} - 0.5| \leq |\lambda_H - 0.5|$ then $L(G\,|\,H',S')$ is as expressible in H' as fit in H or more expressible in H' than $L(G\,|\,H,S)$ in H.

But if $L(G\,|\,H,S)$ is already expressed in H then $L(G\,|\,H',S')$ is also expressed in H'.

Therefore $L(G\,|\,H',S')$ is adapted to, or fits in, H'.

Corollary 2.2. If $\mu(L(G\,|\,H,S), L(G\,|\,H',S')) = 1 - \alpha, \alpha \to 0$ then $L(G\,|\,H,S)$ is as fit in H' as in H. Otherwise $L(G\,|\,H',S')$ is a new language and fitter in H' than in H. Therefore, $L(G\,|\,H',S')$ is an evolution of $L(G\,|\,H,S)$ in H'.

Proof: From Eq. 2.46, if
$$\mu(L(G\,|\,H,S), L(G\,|\,H',S')) = 1 - \alpha, \alpha \to 0$$

then $L(G\,|\,H,S)$ and $L(G\,|\,H',S')$ are the same language, and $L(G\,|\,H,S)$ is therefore adapted to H'. Otherwise $L(G\,|\,H',S')$ is a new language in respect to $L(G\,|\,H,S)$. Therefore $L(G\,|\,H',S')$ is an evolution of $L(G\,|\,H,S)$ in H'.

Theorem 2.3. The fitness $A(L(G\,|\,H))$ of $L(G\,|\,H)$ in a given environment H' is a function of $<\Omega(V_i\,|\,H)>$ of V_n.

Proof: From Eq. 2.23:
$$<\Omega(V_n\,|\,H)> = \Sigma - (<\rho(d(s_i,s_j))> \log <\rho(d(s_i,s_j))>$$
$$- (1 - <\rho(d(s_i,s_j))>) \log (1 - <d(s_i,s_j))>)) V_n$$

such that $<\Omega(V_i\,|\,H)>$ increases if the cardinality of V_n is increased and
$$<\rho(d(s_i,s_j))> \to 0.5.$$

In such a condition, if $\rho(d(s_o,s_t)\,|\,H) \to 1$ is likely that $\rho(d(s_o,s_t)\,|\,H') \to 1$ because
$$\rho(d(s_o,s_t)) = \max \min(d(s_o,s_k) ..., d(s_i,s_j), d(s_m,s_n), d(s_r,s_s) ..., d(s_l,s_t))) V_n$$

and it is likely that here exists $d(s_i,s_j), d(s_m,s_p), d(s_r,s_j) \to 1$ in H and $\rho(d(s_i,s_j),H') \to 1$.

Therefore, the greater $<\Omega(V_i)>$ is, the greater is the number of environments H' for which all (or at least almost all) $\rho(d(s_o,s_t)\,|\,H)) \to 1$ composing $L(G\,|\,H)$ are maintained as $\rho(d(s_o,s_t)\,|\,H')) \to 1$. In such a condition:
$$\mu(L(G\,|\,H,S), L(G\,|\,H',S')) = 1 - \alpha, \alpha \to 0$$

2.6 Knowledge Adaptation and Evolution

Hence, the greater $<\Omega(V_i)>$ is, the greater is the number of environments H' for which $\mathcal{L}(G \mid H, S)$ is adaptable.

Theorem 2. 4. The capacity $E(\mathcal{L}(G \mid H))$ of $\mathcal{L}(G \mid H)$ to evolve in a given environment H' is function of $<\Omega(V_t \mid H)>$ of V_t.

Proof: From Eq. 2.23

$$<\Omega(V_t \mid H)> = -(<\rho(d(s_1, s_t))> \log <\rho(d(s_1, s_t))> -$$

$$(1 - <\rho(d(s_1, s_t))>) \log (1 - <d(s_1, s_t))>)) V_t$$

such that $<\Omega(V_t \mid H)>$ increases if the cardinality of V_t is increased and $<\rho(d(s_i, s_t))> \to 0.5$.

In such a condition, if

$$\rho(d(s_o, s_t) \mid H) \to 1 \text{ and } \rho(d(s_o', s_t') \mid H) \to 0.5$$

it is likely that

$$\rho(d(s_o, s_t) \mid H) \to 0.5 \text{ and } \rho(d(s_o', s_t') \mid H) \to 1$$

and

$$\mu(\mathcal{L}(G \mid H, S), \mathcal{L}(G \mid H', S')) = 1 - \alpha, \alpha \to 0.5.$$

Therefore, $\mathcal{L}(G \mid H', S')$ tends to be an evolution of $\mathcal{L}(G \mid H, S)$.

Hence, the greater $<\Omega(V_t)>$ is, the greater is the number of those environments H' where $\mathcal{L}(G \mid H, S)$ evolves.

Remark 2.13. Theorem 2.2 and its corollary show how a given brain may adapt or evolve a given knowledge $\mathcal{L}(G \mid H, B)$ in response to a changing environment H' by maintaining or improving those restrictions ($\lambda_{H'} \to 1$) upon $G^{\circledR}$, required to keep $\mathcal{L}(G \mid H', B)$ expressible in the new conditions H'. Theorems 2.3 and 2.4 set the conditions in which this occurs. Therefore, the greater the ambiguity of a given genetic grammar, the greater is the genetic plasticity for adaptation to changing environments. Similarly, the conditions for the rejection of $\mathcal{L}(G \mid H, B)$ as an adequate knowledge of new environments are derived from the same theorems. A given $\mathcal{L}(G \mid H, B)$ will be rejected if restrictions imposed by the new conditions H' upon G reduces the expressiveness of $\mathcal{L}(G \mid H, B)$ in H' because $\lambda_{H'} \to 0$ or 1.

2.7 Self-Controlled Expression

Life would be unlikely without the mechanisms of control exerted mainly by the cellular apparatus of enzymatic repair of nucleic acids. Also, protein synthesis is dependent on external sources of both energy and amino acids, and many stps are in charge of controlling the cellular level of these raw materials used to produce other cell components. Therefore, the expressiveness of any genetic language is very much dependent on the mechanism of self-control promoted by a set proteins C over any $d(s_o, s_t)$, that according to definition 2.5 are in charge of either:

Controlling the production or the quantity $q(s_j)$ of proteins $s_j \in d(s_o, s_t)$ that is:

$$q(s_j) = g(q(s_c)), s_c \in C;$$

or, specifying the chemical compatibility $\mu(s_i, s_j)$ among proteins of a given stp, that is

$$\mu(s_i, s_j) = z(q(s_c)), s_i, s_j \in d(s_o, s_t), s_c \in C$$

In this way, the control $Ł(C \mid G, H, S)$ supported by G is exerted by changing $\rho(d(s_i, s_j))$ and therefore by controlling $\Pi(\rho(d(s_i, s_j) \mid H, S))$. This control is dependent on the expressiveness $\theta(Ł(C \mid G, H, S))$ of all those $d(s_o, s_c)$ producing $s_c \in C$ and result in modifying the original S into a controlled space $S^©$. In other words:

$$Ł(C, \mid G, H, S)\, Ł(G \mid H, S) \rightarrow Ł(G \mid H, S^©) \qquad (2.47)$$

The degree of control promoted by $Ł(C \mid G, H, S)$ is dependent on $\theta(Ł(C \mid G, H, S))$.

Theorem 2.5. The control $Ł(C, G \mid H, S)$ maximize $\theta(Ł(G \mid H, S))$ if it promotes

$$\mid \lambda_{H,S}^© - 0.5 \mid \leq \mid \lambda_{H,S} - 0.5 \mid$$

and for all $\rho(d_i(s_o, s_t) \mid H, S^©) \rightarrow 1$ it promotes $\rho(d_j(s_o, s_t) \mid H, S^©) \rightarrow 0$ such that

$$\rho(d_i(s_o, s_t) \mid H, S^©) + \rho(d_j(s_o, s_t) \mid H, S^©) = 1$$

Proof: It is a consequence from theorems **2.1** and **2.2** and their corollaries.

Theorem 2.6. Given $<\Omega(G \mid H, S)>$, $<\Omega(G \mid H', S)>$ as the mean ambiguity of G expressed in H, H', then

$$\mid <\Omega(G \mid H, S)> - <\Omega(G \mid H', S)> \mid \, \propto \, \theta(Ł(C, G \mid H, S))$$

Proof: Given $d(s_o, s_t)$ controlled by $s_c \in C$ then

$$\rho(d(s_l, s_t)) = f(..., q(s_c) ... | H, S)$$

and in the case of a linear control

$$| \rho(d(s_l, s_t) | H, C) - \rho(d(s_l, s_t) | H, C) | \propto q(s_c)$$

But

$$q(s_c) \propto \rho(d(s_o, s_c))$$

and

$$| \rho(d(s_l, s_t) | H, C) - \rho(d(s_l, s_t) | H, C) | \propto \rho(d(s_o, s_c))$$

Also, the number of individually controlled $d(s_o, s_t)$ by $s_c \in C$ is dependent on the cardinality of C. Therefore

$$|<\Omega(G | H, S)> - <\Omega(G | H', S)>| \propto \theta(\mathcal{L}(C, G | H, S))$$

Remark 2.14. Self-controlled languages have better odds of keeping their expressiveness in a greater number of environments than any other replicating grammar G because of their capacity to keep their λ within adequate boundaries. The neuron is a self-controlled grammar processing space, whereas the brain is in charge of controlling the expression of a self-controlled grammar in a multi-organ organism. Because of this, theorems 2.5 and 2.6, combined with theorems 2.3 and 2.4, specify the conditions to improve the success of adaptation and/or evolution of a given knowledge $\mathcal{L}(G | H, B)$. In other words, these theorems set the basic rules of the learning strategies discussed in this book.

2.8 Speeding up Brain Processing

Different alphabets A are used to compose the elements s_i, s_j of $d(s_i, s_j)$ (Eqs. 2.13 to 17 and Remark 2.1). One of these alphabets is supported by ionic currents due to the movement of sodium (Na^+) and potassium (K^+) ions (Fig. 2.4) across the cell membrane. Na^+ is predominant outside of the cell, whereas K^+ is mostly an intracellular ion. These ions move across the membrane through specific channels. The K^+ channel is relatively opened (its conductance g_k is relatively high) and the Na^+ is almost entirely closed (its conductance g_{Na} is almost zero) if the neuron is not activated. The electrochemical equilibrium E_K potential of K^+ is around –90mV and E_{Na} approaches +40 mV. This means that, at rest, the K current is directed to the extra-cellular space and the Na current is toward the inside of the cell. This generates an electrical gradient EM across the membrane due to the different ionic concentrations. EM approaches E_K because the $g_k > g_{Na}$.

The channel state is controlled by both the actual value of the membrane potential EM and molecules composing many stps. Most of the dynamics of this membrane electrochemical ionic system is described by a set of equations proposed by Hodgkin and Huxley (1952) in the first half of the 20th century (Fig. 2.4). These equations describe a dynamic system having two stable equilibrium points (E_K and E_{Na}) and an unstable equilibrium point characterizing a bifurcation (see Fig. 2.4). The actual topology of the ionic system state space is determined by both the number and the states of the ionic channels, i.e., it is under the control of many stps.

On the one hand, it is assumed that g_{Na} is kept very low at the dendrites such that i_{Na} cannot achieve high values. Because of this, whenever a transmitter (s_o) attaches to and opens the Na channel (Fig. 2.4 and Fig. 2.5) it generates a small i_{Na} that promotes a reduction of EM. This EM variation enhances g_K and consequently i_K what tends to counterbalance the i_{Na} enhancement. Because of this, the state point moves around in the E_K surface in the ionic state space. This produces a local variation of EM, called local response s_d, that recodes molecular sign s_o into a electrical sign s_d.

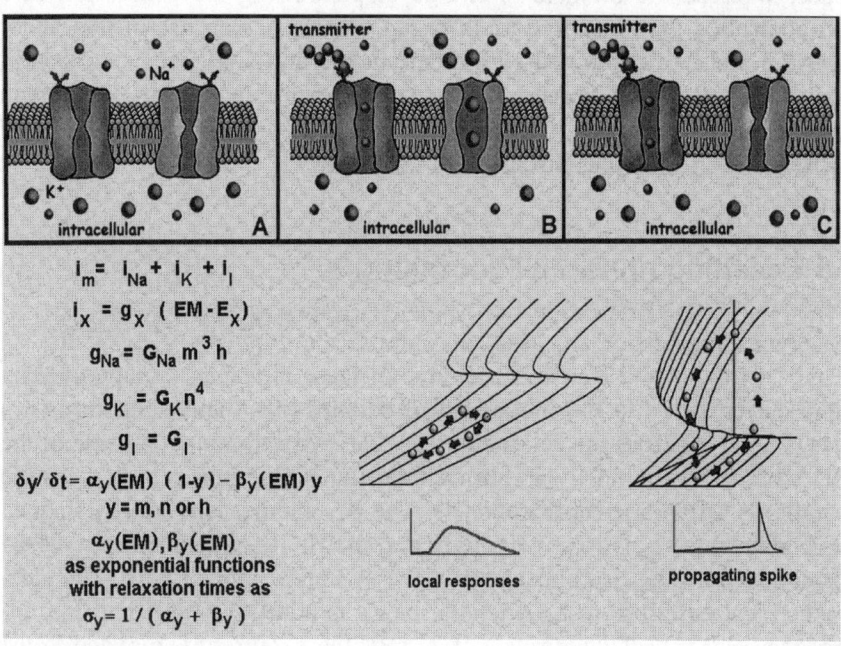

Fig. 2.4. The source of the electrical code

On the other hand, g_{Na} is allowed to reach high values at the axonic membrane, such that now the opening of the Na channel induced by the EM variation at the dendrite may achieve the adequate value for moving the state point to the bifurcation zone. If this unstable zone is reached, the state point jumps from the E_K to the E_{Na} surface and then returns back to E_K. This rapid EM variation is called a spike, and it is strong enough to promote the same dynamic alterations in the nearby Na and K channels.

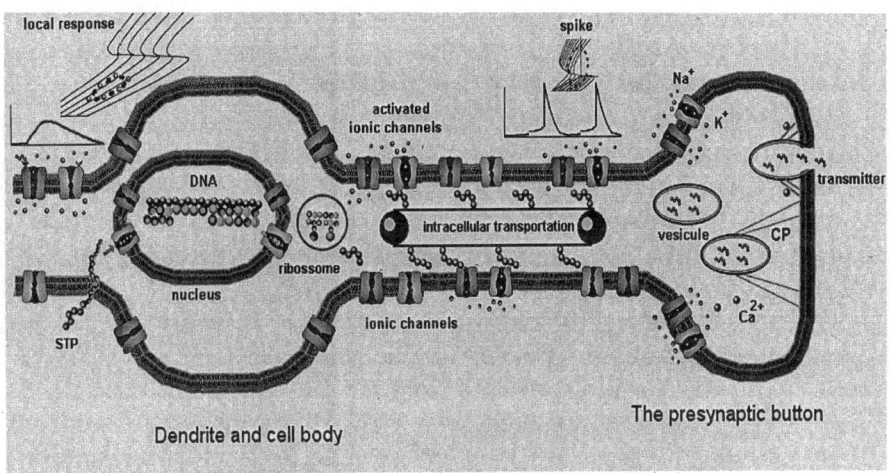

Fig. 2.5. The neuron: *the chemical machinery*

The spike is able to travel along the axon toward its terminal buttons, where the Na and K movements release the calcium ions Ca^{+2} from their storage sites. The released Ca^{+2} activates intracellular contractile proteins (CP in Fig. 2.5) that are able to move the pre-synaptic vesicles toward the cell membrane, in order to release a package of transmitter (s_t) over the next neuron. In this way, the axonic spike train (s_a) recodes the local response s_d and it is recoded by s_t (transmitter) released at the pre-synaptic terminal button. In this way, the derivation chain $d(s_o, s_t)$ describes the neuronal transactions triggered by the release of the transmitter (s_o) by the presynaptic cell, that results into another release of transmitter (s_t) by the postsynpatic cell.

The axonic electrical recoding of the dendritic activity is necessary to speed up chemical message exchange between neurons. The axonic electric code is dependent on the state space topography. Different types of encoding are used by the brain, each one corresponding to a specific state space topography (Rocha, 1997). The movement of molecules in the axon is slow and controlled by the so-called *axonic transportation system*, composed of specialized contractile proteins and activated by means of energy released from special molecules like ATP, which stores chemical energy in phosphate bonds. This axonic transport moves the transmitter or its precursors, produced at the cell body, to the terminal buttons where the transmitters are stored in vesicles. In this way, whenever necessary, chemical information at the dendrites or cell body is recoded into a train of spikes that are quickly transmitted along the axon to trigger the release of chemical information stored at the terminal button (Fig. 2.5).

2.9 The Chemical Talk at the Synapse

Contrary to the simplistic view of a numerically modulated synapse proposed in neural net theory, the complex physiology of the synapse is here formalized by means of a self-controlled grammar G and the expressed languages $L(G \mid H, S_i)$ supported H in each neuronal processing space S_i (Fig. 2.6).

The molecular talk at the synapse is supported by chemical transactions in both directions, i.e., from the pre- to the post-synaptic neuron and vice-versa. Also, the chemicals released from the pre-synaptic neuron may or may not be linked to pre-synaptic spike firing. Many molecules are stored at the pre-synaptic terminal button and released by local molecular processes (e.g., a retrograde signal from the post-synaptic cell: Fig. 2.6), in addition to the transmitters that are released as a consequence of the chemical and electrochemical processing at the pre-synaptic cell body (Fig. 2.5).

Chemicals released from the pre-synaptic cell can act as initial triggers ($d(s_o, s_t)$) of post-synaptic derivation chains $d(s_o, s_t)$ which may or may not involve DNA reading (Fig. 2.6). DNA reading may be part of the control of ionic gates or may be promoted by some other molecule involved in the synaptic machinery. If this is the case, then $d(s_o, s_t)$ may be part of a learning process, as in the line discussed by Rocha (1992), or even in the line of a numerically modulated synapse, as proposed in standard neural nets theory.

However, the final product s_t of $d(s_o, s_t)$ may also be exported (retrograde signal in Fig. 2.6) and used by the pre-synaptic neuron as initial trigger of other pre-synaptic $d(s_o, s_t)$. The retrograde signal may even be transported toward the cell body to control the pre-synaptic DNA reading (Fig. 2.5 and 2.6). Rocha (1997) discussed the role played by this post-synaptic control of the pre-synaptic DNA reading. For instance, dual talk between pre and post-synaptic neurons is very important during the stage of embriogenesis as well as during adult life, in controlling either the neurons' vitality or death (apoptosis).

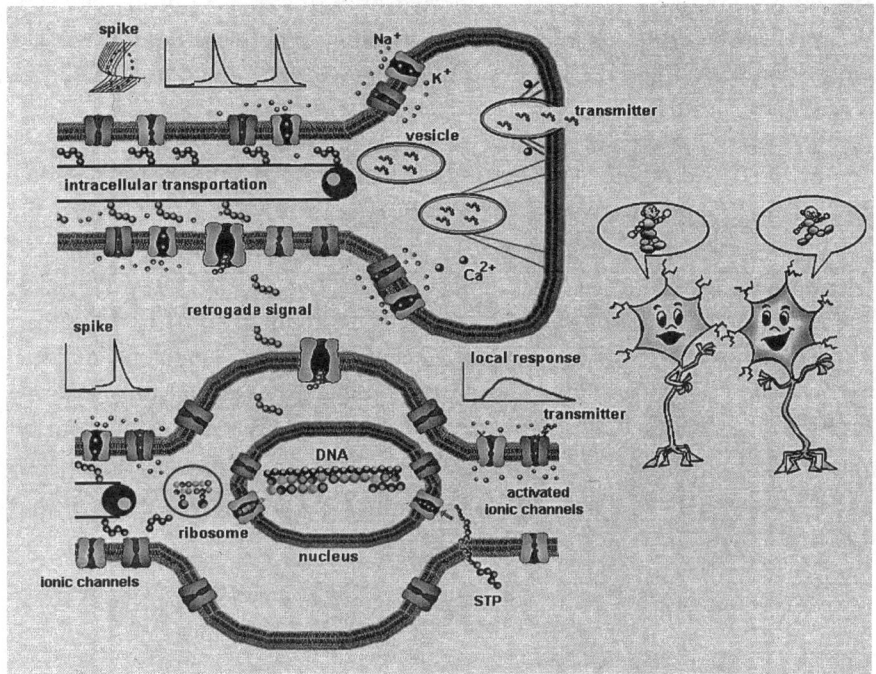

Fig. 2.6. Synaptic chemical talk: *a complex set of chemical transactions take place at the synapse.*

Finally, the exported pre or post-synaptic s_t may be sequestered by other neighboring cells (even glial cells) and may trigger other sets of $d(s_o, s_t)$ in the neighborhood, initiating multiple bilateral or multilateral talks among different types of cells. In this way, the local processing achieves a complexity which depends on the expressiveness $\theta L(G \mid H)$ of G, and the locality of this processing is mainly determined by the temporal restrictions Δ imposed by $<\rho(G \mid H, S^i, S^j, \Delta)>$ in distributing s_t in the neighborhood.

2.10 Summary

The neuron is formalized here as a device handling a subset $Ł(G \mid H, S_i)$ of the formal language $Ł(G \mid H)$, supported by a grammar G in the environment H. Each derivation chain $d(s_o, s_t)$ of $Ł(G \mid H, S_i)$ describes an ordered set of chemical transactions or stp that characterizes a neuronal function. The expressiveness of $d(s_o, s_t)$ is measured by the possibility $\rho(d(s_o, s_t) \mid H, G)$ of their symbol rewritings (or chemical transactions), and it is dependent on both the resource restrictions imposed by H and the self-control $Ł(C, \mid G, H, S)$ allowed by G. The brain B is then formalized here by the set $Ł(G \mid H, B)$ of the languages $Ł(G \mid H, B)$, expressed by each type S_i of neuron composing it. The learnability of any $d(s_o, s_t)$ in $Ł(G \mid H, S_i)$ is, therefore, determined by both the restrictions imposed by H and the total expressiveness $\theta(Ł(C, G \mid H, B)$ of the self-controlling languages of $Ł(G \mid H, B)$. The classical learning procedures used in neural net theories may apply to the neuron described here, for $\rho(d(s_o, s_t) \mid H, G)$ is dependent on the available quantities of the symbols $s_i \in d(s_o, s_t)$. Moreover, new learning procedures may be developed too, exploring the symbolic properties of $Ł(G \mid H, B)$.

3 Brain: A Distributed Intelligent Processing System

The brain is characterized as a Distributed Intelligent Processor of a Distributed Fuzzy Formal Language. This type of modeling takes into consideration recent findings concerning the physiology of the brain, as disclosed by many different brain mapping techniques, such as PET, fMRI, EEG mapping, etc., which allow cognitive functions to be studied in both normal and disabled human beings. The formalization introduced in this chapter provides the theoretical background required for the understanding of the brain as a complex computational device handling numerical, symbolic, and quantum calculations. Each neuron will be considered as a processing space of a subset of derivations supported by a grammar **G**. In this way, each neuron will be considered as a specialized processing subspace of **G** and the result of any cerebral processing will be assumed to be the result of the distributed processing of **G** by a collection of neurons recruited for such a purpose.

3.1 Distributed Intelligent Processing Systems

The following is proposed:

Definition 3.0: A natural or artificial intelligent entity is a system able to efficiently find new solutions to new problems. But the existence of problems suggests both goals and a lack of strategies (plans, models) for efficiently using available tools in order to achieve those goals. To have goals means to have motivation (appetite) to do (obtain) something (e.g., survive). Knowledge of failure in achieving goals demands tools to match performance and goals; matching, in turn, requires memory to store the data used in the evaluation process. Efficiency implies doing a good job under cost and time constraints; effectiveness is evaluated by success in survival. Time and cost efficiency is more likely to be achieved if new solutions may be built up from reorganization and/or generalization of old models — from reselection of tools and capabilities, etc. But whenever necessary, true innovation must be achieved.

Intelligence is a very real property of certain systems, called Distributed Intelligent Processing Systems (**DIPS**), formed by collections of loosely interacting specialized agents. Agents specialize in data collection (sensors), problem solving (experts), data communication (channels), acting upon the surrounding environment (effectors), etc. Intelligence is then approached in terms of a society of communicating specialized experts and the brain is then an example of a natural DIPS (Chandrasekaran, 1981; Davis and Smith, 1983; Ferber, 1999; Fox, 1981; Hewittt and Inman, 1991; Knight, 1997; Lesser, 1991; Maunsell and Ferrera, 1995; Rocha, 1992; 1997, Rocha et al, 2001).

DIPS reasoning is the cooperative activity among an optimally decentralized and loosely coupled collection of experts that may provide the solution of a problem. *Decentralized* means that both control and data are logically and often spatially distributed; there is neither global control nor global data storage. The programming intends to build models in which the control structure emerges as a pattern of passing messages among the agents being modeled. Task distribution is an interactive process between an agent with a task to be executed and a group of other agents that may be able to execute the task.

In this kind of system, intelligence is a function of the types of agents composing the system, but also of the means and purposes with which these agents are used. Intelligence becomes dependent on both the behavior of the specialized agents that are in charge of solving specific tasks and, above all, on the versatility of the relations shared by these specialized agents — or, the plasticity of the commitments to actions among these agents. Of course, the complexity of the tasks solvable by a DIPS determines the number of agents to be enrolled in their solution.

For instance, the task of counting (Rocha and Massad, 2002, 2003a) recruits a set of sensory agents in order to visually inspect the environment and to identify the elements to be counted (Fig. 3.1.). Counting tasks also need to recruit motor control agents in charge of positioning the eyes over the elements to be counted as well as controlling the fingers to point to, or otherwise mark, the identified elements. Also, other agents are in charge of accumulating the number of identified elements and encoding the results into words or numerals. Learning to count, therefore, implies the control of DNA reading in order to create specialized agents for accumulating and classifying data, as well as for controlling the communication resources among these agents at the synaptic level (Rocha and Massad, 2003a). Neuronal specialization is assumed here to be the consequence of the control exerted over $\mathbf{G}^{(d)}$ to define the language $\mathbf{L}(\mathbf{G}^{(d)} \mid \mathbf{H}, \mathbf{n_i})$ to be expressed by the neuron $\mathbf{n_i}$ and learning is assumed to be controlled through specific **stp**s or $\mathbf{d}(\mathbf{s_o}, \mathbf{s_t})$s (Rocha, 1997).

3.2 Distributed Processed Languages

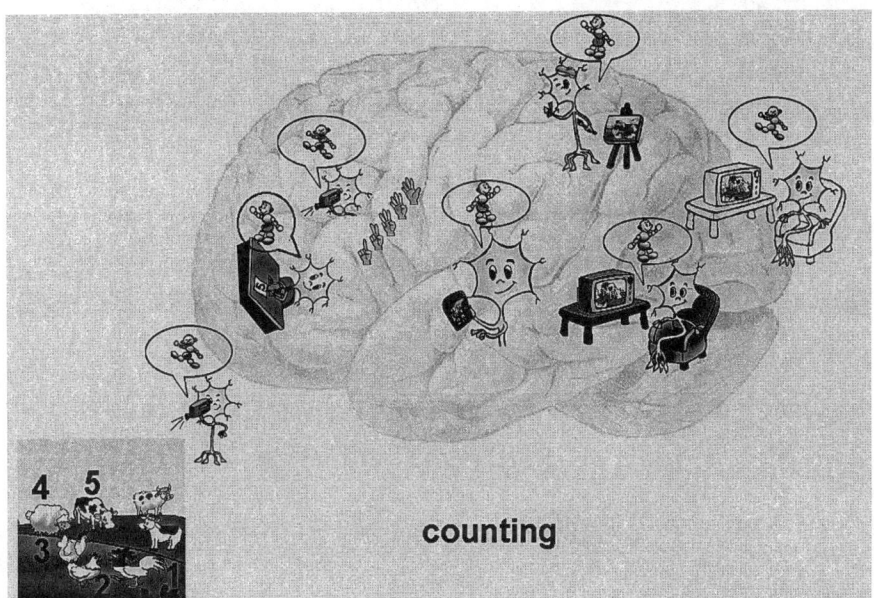

Fig. 3.1. Counting as a distributed task: different neurons take charge of the different tasks in the counting process.

There is one important respect in which our model differs from traditional *neural network* models. In the latter kind of model, all neurons are considered to be identical machines, which differ only in their input-output relations. In our approach, neurons are considered to be internally structured by thousands of **stp**s, formalized by the $d(s_o, s_t)$ supported by a self-controlled grammar $G^{(d)}$. The language $Ł(G^{(d)}(H,S))$ expressed by neurons' cells, as discussed in Chapter 2, specialize them as agents that interact dynamically in a **DIPS**.

3.2 Distributed Processed Languages

In the present context, the set of expressed languages $\{Ł(G(H,S_c))\}_{c=1 \text{ to } n}$ is supported by a grammar $G^{(d)}$.

Definition 3.1: Distributed Processed Grammars (DPG) are a self-controlled grammar of the type

$$G^{(d)} = \{V_0, O \subset D \subset V_n, I \subset V_t, S \subset P, \eta\}$$

where

a) O is a set of genes in charge of specifying the expression of each cellular language $Ł(G^{(d)}(H, S_c))$ of $\{Ł(G^{(d)}(H, S_c))\}_{c=1 \text{ to } n}$ in their corresponding type of processing spaces S_c;
b) I is the set of symbols (inducers) used to specify each $Ł(G^{(d)}(H, S_c))$ such that $d(s_o, s_i | H, S, O, \Delta)$, $s_o \in O$, $s_i \in I$
c) S is a set of rules, such that for $d(s_d, s_t | H, S, \Delta)$ in $Ł(G^{(d)}(H, S_c))$

$$q(s_d) = f(q(s_i)) \text{ and/or } \Delta = f(q(s_i))$$

$$\rho(d(s_d, s_t | H, S, \Delta)) \to 1 \text{ iff } \varphi_m < \Delta < \varphi_n, s_d \in D \qquad (3.1)$$

d) there exist $Ł(G^{(d)}(H, S_i))$, $Ł(G^{(d)}(H, S_j)) \in Ł(G^{(d)}(H, S_c))$ such that

$$V_t(L | S_i, H) \cap V_o(L | S_j, H) \neq \phi; \qquad (3.2)$$

$$V_o(L | S_i, H) \cap V_t(L | S_j, H) \neq \phi$$

Definition 3.2: $G^{(d)}$ is said to be a hierarchical distributed grammar whenever there exists $Ł(G(H, S_b)) \in Ł(G(H, S_c))$ such that

$$V_t(L | S_b, H) \cap V_o(L | S_c, H) \neq \phi$$
$$V_o(L | S_b, H) \cap V_t(L | S_c, H) \neq \phi \qquad (3.3)$$

For any other $Ł(G(H, S_c))$ in $\{Ł(G(H, S_c))\}_{c=1 \text{ to } n}$

Definition 3.3: The organ $O(G^{(d)}(H, S_c))$ is the family of n processing spaces S_c, each $S_c = \{(S_{c,d})\}_{d=1 \text{ to } m}$ composed of m cells c_{cj} in charge of processing a given $Ł(G^{(d)}(H, S_c))$ of a distributed $G^{(d)}$, that is

$$O(G^{(d)}(H, S_c)) = \{Ł(G^{(d)}(H, S_{c,d}))\}_{d=1 \text{ to } mc}\}_{c=1 \text{ to } n} \qquad (3.4)$$

such that

$$V_t(L | H, S_{c,i}) \cap V_o(L | H, S_{c,j}) \neq \phi \qquad (3.5)$$

$$V_o(L | S_{c,i}) \cap V_t(L | H, S_{c,j}) \neq \phi$$

For the sake of simplicity, from now on $G^{(d)}$ will be denoted by G.

3.2 Distributed Processed Languages

The size s of $O(G(H, S^j))$ is its total number of processing subspaces, that is

$$s = \sum_{c=1}^{n} m_c \qquad (3.6)$$

If $\mu(Ł(G(H, S_{c,i})), Ł(G(H, S_{c,j}))) \to 1$ for any i, j then $O(G(H, S_c))$ is called a simple organ, otherwise it is called a complex organ.

Definition 3.4: The multi-organ organism $O(G(H, S^j))$ is composed by a family of organs $O(G(H, S_e))$ that is $O(G(H, S^j)) = \{O(G(H, S_e))\}_{e=1 \text{ to } l}$.

Remark 3.1: The grammar defines a multi-organ organism $O(G(H, S^j))$ and each cell type S_i is in charge of processing a subset $Ł(G(H, S_c))$ of the language $Ł(G(H, S_i))\}_{i=1 \text{ to } n}$ supported by G. In this line of reasoning, each cell of an organ is assumed to be the processing space of a specific set of the chemical language defined by the grammar specified by the genome of the plant or animal. An organ may have just one type of cell, and called a simple organ, or it may have a family of different cells, each one handling a different set of signal transduction pathways for expressing a specific subset of the genetic language. The process of specifying the family of languages $Ł(G(H, S_c))$ of type $O(G(H, S))$ is called embriogenesis, and it is described by the language $Ł(O(H, S))$, supported by $O \subset D \subset V_n$ and $S \subset P$. The genes in O are called *homeobox genes*. These genes are in charge of when and for how long a specific set of genes are activated to build a given organ. The symbols of I correspond to the promoters of early genes E controlled by O. Each homeobox gene controls the build up of a organ or a part of the organism by controlling the expression of the early genes required to specify $Ł(G(H, S_c))$. The set $S \subset P$ is composed by those rules governing the activation of the homeobox genes and the control of these genes upon the early genes.

Theorem 3.1: The mutations of a homeobox gene of O changes the structure of the descendant $O(G(H, S^{j+n}))$ of $O_i(G(H, S^j))$ supported by a given set of expressible languages $\{\{Ł(G(H, S_{c,d}))\}_{d=1 \text{ to } k}\}_{c=1 \text{ to } l}$ either because (1) they change the cardinality of O, modifying the number l of organs of $O(G(H, S^j))$ because they create new processing spaces or cells for a new subset $Ł(G(H, S_{n,d}))$ of the genetic language defined by G; or
(2) because they alter I and consequently change Δ and the size s of a given organ $O(G(H, S_e))$ by changing the number d of a given type of processing space or cell $S_{c,d}$.

Proof: Given any $s_o \in O$ such that

$$d(s_o, s_i \mid H, S, O, \Delta), s_o \in O, s_i \in I$$

$$d(s_d, s_t | H, S), s_d \in D, q(s_d) = f(s_i), s_t \in V_t (Ł (G(H, S_c))$$

a mutation of $s_o \in O$ into $s_o' \in O$ results in

$$d(s_o', s_j | H, S, O, \Delta), s_o \in O, s_j \in I$$
$$d(s_d', s_t' | H, S), s_d' \in D, q(s_d') = f(s_j), s_t' \in V_t (Ł (G(H, S_c))$$

But, if

a) $\mu(s_i, s_j) \to 0$, then it is expected that $\mu(s_d, s_d') \to 0$, such that

$$\mu(S_i, S_j) \to 0,$$

and

1) a new organ $O(G(H, S_e'))$ is created if $s_o \in O, s_o' \in O$ are maintained as expressed genes, then
2) $O(G(H, S_e))$ is removed if $s_o \in O, s_o' \in O$ result in unexpressed genes;
3) otherwise $O(G(H, S_e))$ will be replaced by $O(G(H, S_e'))$;

b) $\mu(s_i, s_j) \to 0,5 < \beta < 1$, then it is expected that $\mu(s_d, s_d') \to \beta < 1$ and

$$\Delta' = f(q(s_j)) \cong \Delta = f(q(s_i))$$
$$\rho(d(s_d, s_t | H, S, \Delta')) \cong \rho(d(s_d, s_t | H, S, \Delta))$$
$$\{Ł(G(H, S_{c,d}))\}_{d=1 \text{ to } k} \cong \{Ł(G(H, S_{c,d}))\}_{i=1 \text{ to } k'}, k \neq k'$$

such that new types of cells may be incorporated into $O(G(H,S_e))$ and/or the size s in Eq. 3.6 is augmented.

Corollary 3.1: Mutations of O belonging to $G(H, S)$ greatly contribute to the creation of new species.

Proof: As a consequence of the fact that any mutation $\mu(s_o, s_o') \to 0$ of O results in changes of the structure of $O(G(H, S))$, by adding, removing or changing a given organ $O(G(H, S_e))$.

Remark 3.2: The meaning of mutation in this book is that of any change of the DNA/RNA nucleotide sequence, during any copying process, due to the grammar ambiguity. The mutation of homeobox genes may result in the genesis of a distinct new species, because it may result in the expression of new sets of languages expressed by G or a huge increase in the processing capacity of a given organ. These are mechanisms that are important in explaining the increases in complexity of the nervous system.

3.3 The Nervous System

Evolution has differentiated animals from other organisms by providing them with a **Nervous System,** or $O(G(H, S_b))$, that assumed the task of controlling the languages $L(G(H,S_c))$ expressed by the other organs of the animal, in order to better adapt it to a changing environment **H**.

Definition 3.5: If $O(G(H, S_b))$ is a complex organ and
$$V_t(L \mid H, S_b) \cap V_o(L \mid H, S_c) \neq \phi \text{ and } V_o(L \mid H, S_b) \cap V_t(L \mid H, S_c) \neq \phi$$
for all other $O(G(H, S_c))$ of $O(G(H, S))$ then $O(G(H, S_b))$ is called the controller (or the Nervous System) of $O(G(H, S))$.

Definition 3.6: The language $L(G(H, S_{b,s}))$ describing the synaptic transactions at $O(G(H, S_b))$ is composed of all expressible
$$d(s_t, s_o \mid G, H, S_{c,i}, S_{d,j})), s_t \in V_t(L \mid H, S_{c,i}), s_o \in V_o(L \mid H, S_{d,j})$$
derivation chains associating two neurons ($n_{c,i}$ and $n_{d,j}$) in charge of processing
$$L(G(H, S_{c,i})), L(G(H, S_{d,j})) \in O(G(H, S_b)),$$
respectively. Thus, for $0 < \lambda < 1$

$$L(G(H, S_{b,s})) = \\ d(s_t, s_o \mid G, H, S_{c,i}, S_{d,j}) \mid \rho(d(s_t, s_o \mid H, S_{c,i}, S_{d,j}, \Delta)) \rightarrow \lambda, \quad (3.7)$$

Remark 3.3: The dynamics of $L(G(H, S_{b,s}))$ are responsible for most of the computational power of the brain as a DIPS because they define the power of the neuronal enrollment in any kind of processing. Also, most of the classical types of learning, such as Hebbian learning, conditioning, etc., supporting connectionist theories involving traditional neural nets, are easily modeled using those G properties that are dependent on the numerical restriction imposed by
$$\pi(\rho(d(s_t, s_o \mid H, S_b)))$$
But new types of learning procedures may profit from the symbolic properties of **G**.

Definition 3.7: $O(G(H, S_b))$ is called an intelligent distributed controlling organ of $G(H, S)$ if
$$<\rho(d(s_t, s_o \mid H, S_b))> \rightarrow \lambda, |\lambda - 0.5| \rightarrow 0$$
for all
$$d(s_t, s_o \mid H, S_b) \in L(G(H, S_{b,s})).$$

Remark 3.4: $L(G(H, S_{b,s}))$ describes both the classically described chemical interaction at the synaptic level, as well as all the chemical transactions between the pre- and post-synaptic neurons required for maintaining the viability of these cells (Rocha, 1997). Also, the conditions imposed in Definition 3.6 guarantee the necessary freedom for any agent to enroll itself with other, different sets of agents in the attempt to solve distinct tasks, and so to render the brain a DIPS.

Theorem 3.2: The augmentation of the cardinality of O increases the types of neurons and/or changes in the composition of I augmenting the number of neurons; are required to increase the complexity of $Ł(G(H,S_b))$.

Proof: As a consequence from Theorem 3.1 and Definition 3.4.

Remark 3.5: The organ in definitions 3.4, 3.5 and 3.6 is a nervous system **NS** that is initially composed of groups of cells distributed over ganglia located at different body sites. Evolution increased the complexity of the NS, and centralized most of the ganglia into the **CNS** (Central Nervous System). As proposed by Rocha (1997) and Rocha et al. (2001) the CNS (or brain) is a DIPS whose intelligence depends on both the type of its neurons and the way these neurons are combined to solve complex tasks. According to Definition 3.6, the brain is an intelligent distributed processor of the expressed language $Ł(G(H, S_b))$ supported by $G(H, S)$ because $Ł(G(H, S_b))$ is dependent on both the specialized neurons processing $Ł(G(H, S_{c,i}))$, $Ł(G(H, S_{d,j}))$ and on $\rho(d(s_t, s_o | G, H, S_{c,i}, S_{d,j}, \Delta)$ S_c that is mostly dependent on the restrictions $\pi(\rho(d(s_t, s_o | G, H, S_b)))$ imposed by H over S_b.

3.4 The Brain

In such a context:

Definition 3.8: The brain is an organ $O(G| H, S_b))$ operating a distributed grammar
$$G = \{S \cup V_o = V_o, V_n, M \cup V_t = V_t, P \mid S \cap V_o = \phi; M \cap V_t = \phi\}$$
where:

a) S is a set of initial symbols sensed in H by means of special sensory spaces S_s, and
b) M is a set of terminal symbols acting over H by means of special motor spaces S_m, such that:
c) $d(s_o, s_t)$, $s_o \in S$ and $s_t \in M$: describes sensory-motor processing;
d) $d(s_o, s_t)$, $s_o \in V_o$ and $s_t \in V_t$, describe internal processing;
e) $s_o \in V_o$ and $s_t \in M$ then $d(s_o, s_t)$ describes volitional action over H similar to $d(s_o, s_t)$; and if

f) $s_o \in S$ and $s_t \in V_t$ then $d(s_o, s_t)$ describes a memory about H, similar to $d(s_o, s_t)$ of the organism $O(G(H, S))$;

Remark 3.6: In this context, the brain is viewed as a distributed processor of the genetic grammar defined by G, composed of at least three superfamilies of neurons, namely *Sensory Systems, Effector* (mainly *Motor*) *Systems* and *Internal Processing Systems*. Many different types of neurons n will be part of these superfamilies, each one of them characterized by a given subset $\mathcal{L}(G(H, S_n))$ of all expressible languages supported by G in H.

Definition 3.9: The knowledge **K** of $O(G(H, S))$ about H is

$$K = \{d(s_o, s_t) \mid d(s_i, s_j) \subset d(s_o, s_t) \cap d(s_o, s_t) \neq \phi\} \quad (3.8)$$

Remark 3.7: Each item of knowledge K is then shared by $O(G(H, S_b))$ considered here to be described by a subset $\mathcal{L}(G(H, S_k))$ of the language supported by G in H that have a non-empty intersection with the sensory-motor language $\mathcal{L}(G(H, S_m)) = \{d(s_o, s_t) \mid s_o \in S \text{ and } s_t \in M\}$ promoting the survival of $O(G(H, S))$ in H.

Definition 3.10: *Learning by observing* is the process of acquiring knowledge $d(s_o, s_t) \in K$ about H from the observation $d(s_o, s_t)$ of H.

Remark 3.8: This type of learning is therefore the result of the processes that modify the expression of the languages used by the neurons of a given $O(G(H, S_b))$. This may be achieved either by changing the neuronal enrollment by modifying $\rho(d(s_t, s_o \mid H, S_{c,i}, S_{d,j}, \Delta))$, or changing the language $\mathcal{L}(G(H, S_{k,n}))$ expressed by a set of neurons $S_{k,n}$, in order to create a new specialized agent.

Proposition 3.1: DIPS reasoning involves different types of agents because it entails:

a) *defining a goal to be achieved*: selecting a need to be fulfilled implies the existence of agents to detect the system's actual needs;
b) *determining the DIPS' actual state:* performing a global evaluation of the main ongoing activities, implying the existence of agents to monitor the agent enrollments supporting those ongoing activities;
c) *planning the means of achieving goals:* selecting and organizing available tools judged to be adequate for solving problems, which implies the existence of agents able to recruit and organize other agents based on the actual state and past experience;
d) *retaining useful information:* recalling past experiences in order to orient planning and to store data about ongoing planned activities,

implying the existence of agents specialized to memorize such pieces of information;
e) *utilizing a set of disparate tools:* developing tools adequate to different tasks which may be modified or combined for new purposes, implying the existence of different agents specialized in handling or modify tools or creating new ones;
f) *evaluating the performance in achieving the desired goal:* comparing what is being done with what was planned, implying the existence of agents able to evaluate the progress (or lack thereof) of the planned action;
g) *optimizing communication:* promoting message exchange between all the agents to (1) announce the task; (2) support its solution by those enrolled agents; and (3) alter enrollment if necessary. All of this implies the capability of each agent to relocate its communication resources whenever necessary.

Proof: It follows from definitions 3.0 and 3.7.

Remark 3.9: Reasoning supported by $O(G \mid H, S_b))$ involves different types of neurons (Fig. 3.2) in charge of:

a) *goal definition:* or, detecting a need to be fulfilled, such as when the limbic neurons detect basic necessities like food, water, etc., or when frontal cells guide environmental searches;
b) *attention control:* or, monitoring actual agent activities, such as when frontal and parietal cells control limbic and sub-cortical agents involved with the control of cortical activity;
c) *planning:* or, selecting and organizing available agents judged to be adequate to solve the problem, as for example frontal recruiting agents described in the literature as executive agents (see also Chap. 5);
d) *memorizing useful information:* or, encoding and recalling both retrospective (hypocampus) and prospective memories (frontal lobe);
e) *implementing tools:* or, monitoring (sensory systems) and acting (motor systems) upon the environment, storing knowledge (e.g. in semantic memory), etc;
f) *evaluating performance:* or, verifying if needs are being fulfilled, via the emotional agents located in the limbic system and the frontal neurons which recruit them; and
g) *communicating resource information*: as provided by transactions supported by $\mathcal{L}(G(H, S_{b,s}))$ (see definition 3.6).

Definition 3.11: $G(g) = \{V_g, V_n, V_e, P\} \subset G$ is the grammar supporting the processing of the goal **g** where:

a) $\mathcal{L}(G(H, S_g))$ is the language expressed by the agents S_g in charge of defining the goals g of $O(G \mid H, S_b)$);
b) $\mathcal{L}(G(H, S_e))$ is the language expressed by the agents S_e in charge of verifying if the goals g of $O(G \mid H, S_b)$) are being fulfilled;
c) $V_g = \{s_t \in d(s_0, s_t) \mid [\, d(s_0, s_t) \in \mathcal{L}(G(H, S_g)), \rho(d(s_0, s_t)) \to 1]\}$ and
d) $V_e = \{s_t \in d(s_0, s_t) \mid [d(s_0, s_t) \in \mathcal{L}(G(H, S_e)), \rho(d(s_0, s_t)) \to 1]\}$

In this context:

Proposition 3.2: The DIPS reasoning $R(G(g) \mid H, S_b))$ about g is
$R(G(g) \mid H, S_b)) =$
$\quad \{d(s_0, s_t) \mid [\rho(d(s_0, s_t))) \to 1 \wedge s_0 \in V_g, s_t \in \mathcal{L}(G(H, S_e)]\}$
Also:
a) acceptance V_g of such a reasoning is
$$V_g = \{s_t \in \mathcal{L}(G(H, S_e)) \mid \rho(d(s_0, s_t))) \to 1\};$$
b) refutation $\sim V_g$ of such a reasoning is
$$\sim V_g = \{s_t \in \mathcal{L}(G(H, S_e)) \mid \rho(d(s_0, s_t))) \to 0\};$$
c) degree of acceptance $\mu(R(G(g) \mid H, S_b)))$ of $R(G(g) \mid H, S_b))$ as a solution for g is
$$\mu(R(G(g) \mid H, S_b))) = (V_g \cap V_g) / V_g$$
d) degree of refutation $\sim \mu(R(G(g) \mid H, S_b)))$ of $(R(G(g) \mid H, S_b))$ as a solution for g is
$$\sim \mu(R(G(g) \mid H, S_b))) = (\sim V_g \cap V_g) / V_g.$$

Proof: follows from proposition 3.1 and definition 3.11.

Remark 3.10: From Definition 3.9 & Proposition 3.2, $(R(G(g) \mid H, S_b))$ is supported by the knowledge

$$K = \{d(s_0, s_t) \mid [\rho(d(s_0, s_t))) \to 1 \text{ and } s_0 \in V_g, s_t \in V_g]\}.$$

Let this be denoted by $R(G(g) \mid H, K, S_b))$.

64   3 Brain: A Distributed Intelligent Processing System

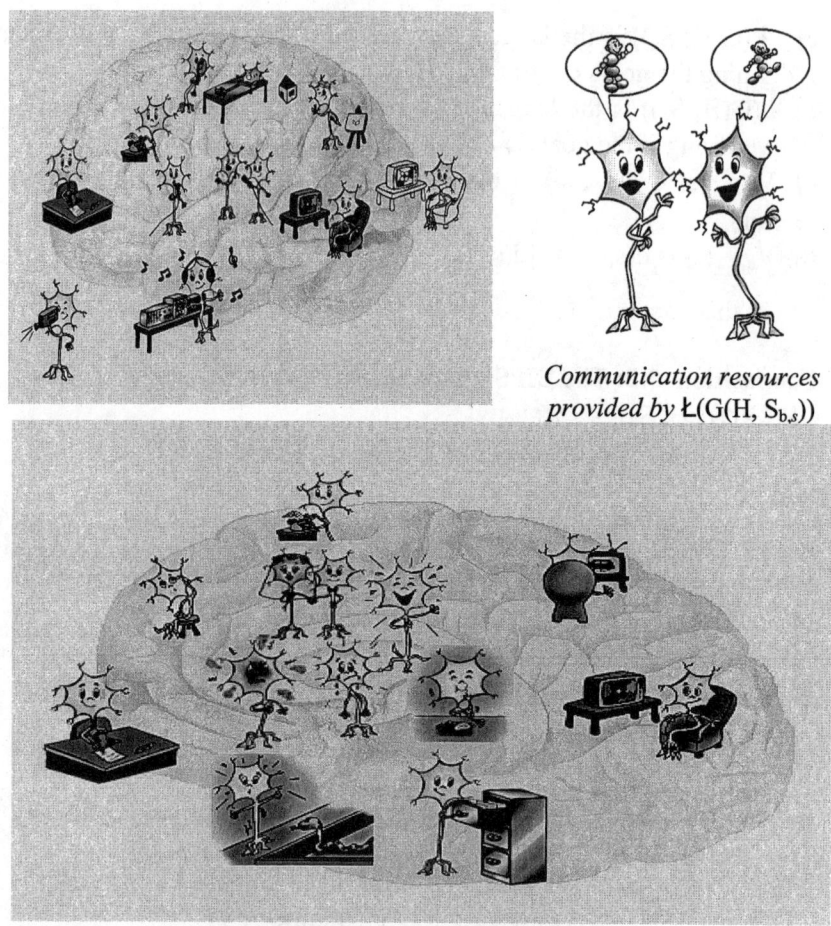

Communication resources provided by $Ł(G(H, S_{b,s}))$

Agent specialization supported by $Ł(G(H, S_c))$

Fig. 3.2. Neural agents supporting reasoning: *reasoning is the result of the interaction among a complex set of specialized neurons.*

3.5 Brain Communication Channels

Communication among DIPS' agents is established by means of two main strategies:

a) *Mail addressing*: both the sending and the receiving agents know themselves; that is to say they have the capacity to address messages specifically to each other. It its the case of the many neural tracts that specifically innervate defined cerebral areas or nuclei; and

b) *Blackboard posting:* agents deliver messages that are not specifically addressed to another defined agent, but to those interested in the subject. This is clear in the case of hormones released in the blood stream, which act upon cells whose physiologies require these hormones. It is also the case of the neuromodulators that are broadly released over large areas of the brain by means of a very spread axonic net. Neuromodulators have different binding (sites where matching occurs) and effector (rewriting) sites. It is also the case of those broad neural circuits controlling brain reactivity, as in the case of the waking, sleep and arousal states. These distinct communication strategies play different roles in brain processing and learning:

Hormones are assumed to be general messages released by a group of agents and broadcasted by means of the blood stream to influence many parallel and independent processes in order to coordinate their actions. It may be a useful process to broadly spread information about the system's state and/or goals to be achieved;

Neuromodulators or neuropeptides are used to coordinate the activities of neural systems operating in a parallel fashion and/or according to some general hierarchy. Modulators will also be very useful in controlling learning by means of many different strategies, because they may be used to control synaptic specification, growth, stabilization and/or death. For instance, they may be used to promote the growth of the synapses among well succeeded agents and to reduce communication resources among neurons that cooperated in a failed attempt to reach a specified goal; and

Neurotransmitters are used either as general messages broadly released in the brain by means of very spread neural circuits (e.g. those related with arousal and sleep control) or as local information in those more specific neural tracts (e.g., cortical-spinal and spinal-cortical circuits controlling the muscles). In the first case, neurotransmitters will set the general operating conditions for defined groups of agents, such as specifying the type of axonic code to be used; the set of useful messages, etc. In the second case, neurotransmitters will be part of the actual processing and because of this they will be under the control of learning mechanisms related with the plasticity of the commitment of agents in solving tasks. Changes in the amount of communication resources (e.g., transmitter) modify the capacity of any agent to enroll in the solution of a given task.

3.6 The Basics of DIPS Learning

According to Definition 3.0, learning is the tool used by any intelligent system to solve the problem when a defined goal **g** is not achieved. This section is devoted to discussing this subject.

Theorem 3.3: $R(G(g) | H, K, S_b))$ has to be reviewed whenever
$$\mu(V_g, V_g | H, S_b) \rightarrow .5.$$
Proof: As consequence from the fact that the difficulty in satisfying g increases as
$$\mu(V_g, V_g | H, S_b) \rightarrow 0.5.$$
Corollary 3.3: The learning of a new $R(G(g) | H, K, S_b))$ is necessary if
$$\mu(V_g, V_g | H, S_b) \rightarrow .5 \text{ for and given } R(G(g) | H, K_o, S_b))$$
Proof: As a consequence of the fact that $R(G(g) | H, K_o, S_b))$ does not provide a solution for g as $\mu(V_g, V_g | H, S_b) \rightarrow .5$.

Theorem 3.4: If $R(G(g) | H, K_o, S_b))$ has to be reviewed, then DIPS learning modifies $\Pi(\rho(d(s_i, s_j) | H, S_b)))$ by changing $\rho(d(s_i, s_j) | H, S_b)$ and/or creating new $s_n \in V_n \cup V_g$ in order to guarantee a new
$$K = \{d(s_0, s_t) | [(\rho(d(s_0, s_t)) \rightarrow 1) \text{ and } (s_0 \in V_g, s_t \in V_g)]\},$$
such that $\mu(V_g, V_g | H, S_b) \rightarrow 1$.

Proof: Let it be supposed that a given knowledge K_o, developed under $\Pi_o(\rho(d(s_i, s_j) | H, S_b)))$, supports a given reasoning $R_o(G(g) | H, K_o, S_b))$ as an attempt to achieve the goal g. Also, let it be assumed that
$$\mu(V_g, V_g | H, S_b) \rightarrow 0.5.$$

In this condition

$$V_g = \{s_t \in d(s_0, s_t) | d(s_0, s_t) \in \mathcal{L}(G(H, S_g)) | \rho(d(s_0, s_t)) > .5 + \xi, \xi \rightarrow 0\}$$

for some $d(s_0, s_t)$, because $\rho(d(s_i, s_j)) \rightarrow .5 + \xi$, for $d(s_i, s_j) \subset d(s_0, s_t)$. Now, from Definition 2.1:

$$\rho(d(s_i, s_j)) = \Psi_{k=1}^{n}(f(q(s_i), q(s_k), q(s_j), \mu(s_i, s_k) | H)) \rightarrow 1$$

But, from Definition 2.5 for a self-controlled grammar and Eq. 2.34 defining such a control

$$q(s_i) \text{ and/or } q(s_k) \text{ and/or } q(s_j) = g(q(s_c)) \text{ and/or } \mu(s_i, s_k) = g(q(s_c))$$

3.6 The Basics of DIPS Learning 67

Now, if $q(s_C)$ is also a function of $\mu(V_g, V_g \mid H, S_b)$, then it is possible to change $\rho(d(s_i, s_j) \mid H, S_b)$ to increase $\mu(V_g, V_g \mid H, S_b)$. This implies that reorganizing knowledge K_o into K to update $R_0(G(g) \mid H, K_o, S_b))$, and this is done by changing $\rho(d(s_0, s_t))$.

If s_n in $d(s_d, s_n)$ is an unexpressed symbol of G then $q(s_n) = 0$ because $\rho(d(s_0, s_n \mid H, S_b)) = 0$. But according to Eq. 2.47 it is possible to implement a control

$$\mathcal{L}(G \mid H, S_b) \xrightarrow{\text{not } K_o} \mathcal{L}(G \mid H, S_b^{\copyright})$$

to obtain

$$\rho(d(s_0, s_n \mid H, S_b^{\copyright})) > 0 \text{ and } q(s_n) > 0$$

and s_n is expressed and incorporated into $V_n \cup V_g$.

In these conditions, let

$$\Pi_0(\rho(d(s_i, s_j) \mid H, S_b))) \xrightarrow{\text{not } K_o} \Pi(\rho(d(s_i, s_j) \mid H, S_b))),$$

then a new

$$K = \{d(s_0, s_t) \mid [\rho(d(s_0, s_t)) \to 1] \text{ and } [s_0 \in V_n \cup V_g, s_t \in V_g]$$

is developed under $\Pi_0(\rho(d(s_i, s_j) \mid H, S_b)))$ to support $R(G(g) \mid H, K, S_b))$.

Theorem 3.5: The learning capability of a DIPS is determined by the ambiguity

$$<\Omega(G \mid H, S_b)> \text{ of } G(H, S_b)$$

and by the control capability

$$\theta(L(C \mid H, S_b)) \text{ of } \mathcal{L}(C \mid G, H, S_b).$$

Proof: On the one hand, the expressiveness

$$\theta(L(C \mid H, S_b)) \text{ of } \mathcal{L}(C \mid G, H, S_b)$$

determines how much control over $\Pi_0(\rho(d(s_i, s_j) \mid H, S_b)))$ may be implemented according to Eqs. 2.34 and 2.47, and this limits the changes that can be promoted over any $\rho(d(s_i, s_j))$ supported by $G(H, S_b)$, whether s_i belongs to $D \subset V_n$ or not.

On the other hand, the ambiguity of $<\Omega(G \mid H, S_b)>$ determines the capacity for turning on the expression of $d(s_d, s_n)$ to generate a new expression $s_n \in V_n \cup V_g$.

Therefore, both $< \Omega(G \mid H, S_b) >$ and $\theta(L(C \mid H, S_b))$ determines what is learnable by a given $O(G \mid H, S_b))$.

Remark 3.11: Theorem 3.4 specifies two different conditions for modifying a given knowledge K. The first one changes the association among a defined set of specialized agents because it implies modifying the possibility of the already expressed $d(s_i, s_j)$s, whereas the second strategy implies creating a new specialization because it will change the set of symbols V^*. The first strategy ($K_0 \rightarrow K_{0'}$) is more a process of optimizing or adapting a pre-existent knowledge K_0 to new conditions. It is a sort of knowledge re-engineering. The second approach ($K_0 \rightarrow K_1$) is more a process of knowledge evolution by means of which brand new solutions to new problems arise. The change of V^* may be achieved by merely promoting the expression of an already existing gene induced by the changing environment, which poses new questions, or by means of a mutation creating a new gene to bring about a new $s_n \in V_n \cup V_g$. Mutation requires at least one generation to be accomplished because it has to occur first at the level of gametes to be later expressed at the level of somatic (neural) cells. However, mutation may also be hypothesized to occur during mitosis, creating a new cell, most probably during embriogenesis or even during post-natal life. In the first case, different beings sharing the same genetics will have different learning capabilities, although such differences cannot be large as it is not expected that the mutation rate is high. In the second case, very new specialization may result from demands of the changing environment that must first of all promote the creation of new cells. The process can be improved if the mutation rate is increased in such a condition, to guarantee that these new cells S_n will express new languages $L(G \mid H, S_n)$). This may be the case if some genes $s_n \in D$ are specially designed to generate high V^* variability, as has been discovered in the cases of genes governing the immune cells or chemical receptors at the olfactory cells and recently proposed to be also at work in the brain (Arshavsky, 2002). In any case, when new cells are created in post-natal life, the new knowledge K_1 has to be rediscovered by each individual or has to be copied by epigenetic means from those brains which first attained it. This is the role played by memetics.

A clear example of knowledge evolution is the human transformation of fuzzy quantification (K_0) capability, which is shared with the animals, into the crisp numbers (K_1) supporting modern arithmetic (Fig. 1). Our key proposition in this book (see Chap. 4) is that the creation of crisp numbers is achieved by changing the specialization of some cells in the brain, which may be accomplished by changing the expression of some genes already existing in $D \subset V_n$ of many animals and in charge of defining certain ionic channels and some topographic molecular markers guiding axonic growth. This proposition is supported by the fact that neurons of some primates are specialized in identifying numbers above 5 (see Chaps. 1 and 7) and by the fact that man invented crisp numbers on at least in four different occasions

(Fig. 3.3) in four non-overlapping civilizations (e.g., Ifrah, 1985, Joseph, 1990).

Fig. 3.3. Number, evolution, and neural circuits: *whenever culture pushed, the brain improved its number of neural circuits.*

It will be assumed here that demands imposed by increasing trade activities have pushed the expression of such genes, whenever the complexity of the human society required it. This could explain why our Tupinambás (Fig. 1.1) were still using primitive number systems when the Portuguese people arrived in Brasil, as the result of a huge effort to increase their commercial trade with India. Once crisp numbers are created by a human culture they may be transmitted from generation to generation of this culture or to others by means of memetic reproduction as discussed in Chap. 6. This is because possibility cannot be ruled out, here, that the evolution of the mathematical human capability is taking profit of meme-gene co-evolution to create new $s_n \in D \subset V_n$.

3.7 Evolutionary Learning

To learn is to model the observable world in order to understand it (Ferber, 1999, Rocha, 1982 a, b). A model is a set of relations between data or evidence obtained with a set of instruments, and actions performed by a set of acting (motor) agents. Understanding requires that the model to fulfill some defined purpose (or goal), which may be a simple survival task, or complex intellectual processing. The goal is set because some agents detect some need for resources or data to accomplish a defined task. From a general point of view, to understand is, therefore, to provide a set of adequate responses in order to adapt the system to the surrounding world, or in other words, to maintain its identity in a changing environment. In such a line of reasoning:

Definition 3.12: Modeling is characterized as:

a) *Detecting a motivation to act:* the set H of agents in charge of monitoring the actual conditions of the DIPS agents, detect a necessity $q(s_n)$ of resource or data to support ongoing activity

$$d(s_o, s_t) = \alpha\, s_o\, \beta \rightarrow ... \rightarrow \alpha\, s_n\, \beta\, ... \rightarrow \alpha\, s_t\, \beta.$$

In this context, the set N of necessities evaluated by H is:

$$\{q(s_n) \text{ of } s_n \in V^{\#} | (\alpha s_o \beta ... \rightarrow \alpha s_n \beta \rightarrow \alpha s_{n+1}\beta ... \alpha s_t \beta)\}, \text{ and}$$

$$\rho(s_n, s_{n+1}) \rightarrow 0;$$

b) *Setting a goal g definable over H:* another set G of agents uses information about N to select a set of final states given by $V_g \subset V_t$ according to definition 3.11 and proposition 3.2, as those states to be attained in order to satisfy the query posed by N. Thus:

$$g: N \times V_t \rightarrow (0,1) \text{ and } g: N \rightarrow V_g$$

Once V_g is determined, it is necessary to define the set of $d(s_o, s_t)$ supported by the grammar $G(g) = \{V_n, V_g, P\}$ that fulfill g because $\rho(d(s_o, s_t)) \rightarrow 1$, $s_o \in V_n$, $s_t \in V_g$. Thus, according to proposition 3.2, this is accomplished by

$$R(G(g)|H, K, S_b)) = \{d(s_o, s_t) \mid [\rho(d(s_o, s_t)) \rightarrow 1] s_o \in V_g, s_t \in V_g;$$

c) *Setting the relevance $\tau(s_t)$ of each $s_t \in V_g$:* since V_g is a fuzzy set, not all s_t are supposed to have the same relevance to satisfy N; The actual value of $\tau(s_t)$ is set as a function of the amount $q(s_n)$ of resource or data needed. Thus:

$$\tau(s_t) = f(q(s_n)).$$

3.7 Evolutionary Learning

This is the task of the set H of agents in charge of evaluating the attainability of g in definition 3.11;

d) *Collecting a set F of facts, evidences or data about g from H with a set of sensory agents S:* this process S of sensory information collecting is: $S : S \times H \times F \times g \rightarrow (0,1)$ that is, obtaining a set F of measures about H with the set of sensory agents S. The set of evidences F can always be described by the language $\mathcal{L}\,(G \mid H, S)$, that is $S : S \times H \times g \rightarrow \mathcal{L}\,(G \mid H, S)$. In general, F is a redundant set of measurements performed by S. In other words, there are many similar $d(s_o, s_t) \in (G \mid H, S)$ in F. Let $\mathcal{F}$ be the non-redundant set of evidence associated to F. Thus $\mathcal{F} \subset F$;

e) *Recognizing the sensory images $I(o_i)$ of objects o_i of interest in H as sets of relations between the collected pieces of evidence or facts.* The process of sensory recognition R is to obtain the set of relations $I(o_i)$ among the measurements in the non-redundant E associated to an object o_i of H. To recognize R is

$$R: (\mathcal{F})^n \times I(o_i) \times g \rightarrow (0,1)$$

Recognition, therefore, consists in the result of the calculations performed by a set of agents C specialized in classifying the objects $o_i \in U$ according to the set $I(o)$ of relations between the measures m_i about these o_i. $I(o_i)$ is therefore described by $\mathcal{L}\,(G \mid H, R)$:

$$R: (\mathcal{F})^n \times g \rightarrow \mathcal{L}\,(G \mid H, C);$$

f) *Analyzing the possible behavior B(o) of the identified objects $I(o_i)$:* this analysis of **B** is performed by a set of agents **B** having the adequate set of tools for such an analysis:

$$B: I(o) \times g \times B(o) \rightarrow [0,1] \text{ or } B: I(o) \times g \rightarrow \mathcal{L}\,(G \mid H, B);$$

g) *Planning P(g) how to reach g given B(o) and the available tools T:* that is, to assess how attainable g is from $I(o_i)$ given resources T. To plan is $P : B(o_i) \times g \times P \times T \rightarrow (0,1)$, and it is the result of the calculations performed by a collection of agents P using defined rules. The possible plans are provided by $\mathcal{L}\,(G \mid H, P)$. Thus: $P: B(o) \times g \rightarrow \mathcal{L}\,(G \mid H, T);$

h) *Making a decision to act A(g) over H using a set of (motor) agents M:* this decision D is based upon P about the selection of the best agents to act over H in order to achieve g and is the duty of a set of specialized agents D. To make a decision is

$$A : P(g) \times D \times M \rightarrow (0,1) \text{ or } A : P(g) \times D \rightarrow \mathcal{L}(G \mid H, M)$$
$$\mathcal{L}(G \mid H, M) = \{d(s_o, s_t) \mid [s_o \in V_g, s_t \in M\,]\}$$

i) *Evaluating $E(g)$ if the executed actions $A(g)$ fulfilled the desired g:*
the executed actions $d(s_0, s_t)$ are supposed to result in the required $q(s_n)$ of s_n defining $s_t \in V_g$. Thus $E: A \times H \times g \to (0,1)$ such that $d(s_0, s_t) \to d(s_t, s_n)$, $d(s_0, s_t) \in R(G(g) | H, K, S_b))$, and:
$$d(s_t, s_n) \in \mathcal{L}(G | H, E)\ E: A \times H \to V_g$$
$$V_g = \{s_n \in V_g | \rho(d(s_t, s_n)) \to 1\}$$
In this way, it is possible to calculate both the model's *acceptance* and *refutation* (definition **3.2**) as
$$\mu(R(G(g)) | H, K, S_b)) = (V_g \cap V_g) / V_g, \sim\mu(R(G(g)) | H, S_b))$$
$$= (\sim V_g \cap V_g) / V_g)$$
respectively. Such an evaluation is the duty of a set of specialized agents E, such that: $E : V_g \times V_g \to \mathcal{L}(G | H, E)$

i) *The attainability of* $s_t \in V_g$ is encoded by two strings $d(s_t, s_r)$, $d(s_t, s_p)$ produced by two different sets evaluating sets E_r, E_p of agents calculating the reward (s_r) and punishment (s_p) of s_t as belonging to V_g respectively, such that given ° as a T-norm
$$\mathcal{L}(G | H, E_r) = \{d(s_t, s_r) | [\rho(d(s_t, s_r)) = \rho(d(s_t, s_n)) \circ \tau(s_t)]\}$$
$$\mathcal{L}(G | H, E_p) = \{d(s_t, s_p) | [\rho(d(s_t, s_p)) = ((1 - \rho(d(s_t, s_n))) \circ \tau(s_t)]\}$$
In this condition, the adequacy $\mu(M_0(g) | H, K, S_b)$ of $M_0(g)$ to fulfill the goal g given H, K, S_b is
$$\mu(M_0(g) | H, K, S_b) = \rho(d(s_t, s_p)) - \rho(d(s_t, s_r))$$
In such a condition:

φ) *$M_0(g)$ adapts $O(G | H, S_b)$ to H according to g if*
$$\mu(M_0(g) | H, K, S_b) \to 1$$

and $M_0(g)$ has to be a rejected as solution of g in H if
$$\mu(M_0(g) | H, K, S_b) \to -1.$$

Remark 3.12: The necessities of the DIPS agents are of two kinds: tool maintenance and data processing (Fig. 3.4). The first type of necessity is directly related with the house keeping of each agent. For instance, in the case of the neuron, the cell has to deal with energy and structural component supplies to keep itself alive. The second kind of necessity is directly related with the activities the agent is involved in. Whenever it accepts a task the cell may have need of specific types of data. In the case of neurons, this may imply sensing the environment in specific searches.

The first kind of necessity is in general periodic and most of the structure of the models M_s required to accomplish survival tasks are hardwired or learned in the first stages of DIPS development. In the case of animals, these tasks are related to feeding, sheltering, looking for sexual partners, taking care of offspring, etc., and the neural circuits implementing such tasks are mostly organized at the level of the limbic system involve special neurons or receptors to monitor many body parameters like blood pressure, temperature, oxygen supply, sugar, salt, etc. Or they may be involved in organizing sexual, parental, and social behaviors, etc.

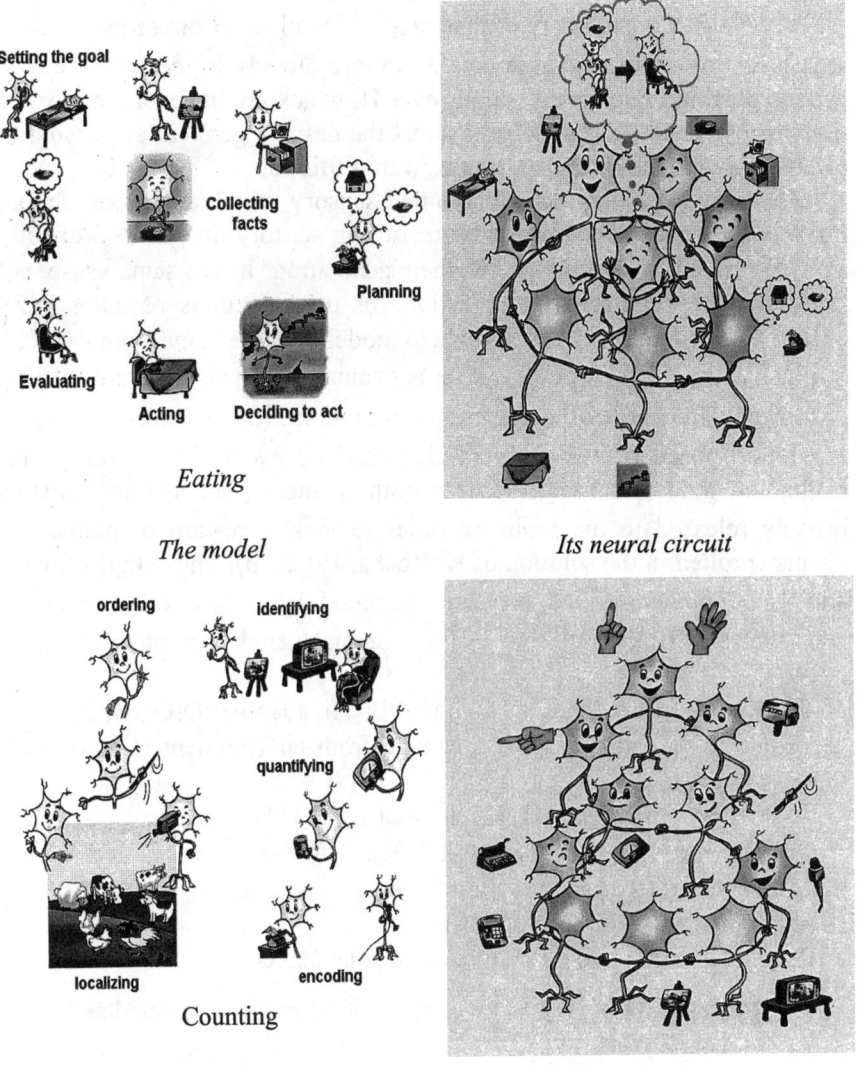

Fig. 4. Modeling by DIPS: *to create a model implies having a set of specialized agents able to implement it.*

The second type of necessity is more sporadic and variable, because it is mostly determined by the exploration of H and the attempts to make the survival activities easier. The neural circuits implementing such tasks are mostly organized at the neocortex level and involve many different types of specialized cells to sense or manipulate the environment. Once a necessity is detected, memory has to be scanned to provide information about previously successful and unsuccessful actions, or about places, means, etc., of fulfilling such a necessity. All these kind of information are used to set each $s_t \in V_g$ and its relevance $\tau(s_t)$, as well as to direct the sensory search for information. These data and those already in memory are then used to plan and implement actions over H, which are in general motor actions whose results are matched against the defined goals in an attempt to evaluate how much of the necessities were fulfilled.

Models may or may not involve real sensory search and motor action. Sensory data may be simulated by recruiting sensory images $I(o_i)$ directly from R instead of involving S in their generation. In the same sense, actions may be simulated motor actions or manipulations of other DIPS agents. Because of this, it is possible to model imagined environments H.

Finally, the model's performance is evaluated by two different sets E_p, E_r of agents in charge of evaluating its acceptance (E_r) or rejection (E_p) as a solution posed by the problem characterized by N. The result of this evaluation guides learning, because both s_r and s_p are neuromodulators broadly released in the brain, in order to locally reward or punish the agents enrolled in the solution of N (Rocha, 1982a, b). The details of the s_r and s_p are locally defined, because neuromodulators have different action sites that are locally selected by the chemical environment of the local agent.

Theorem 3.6: Let there be a model $M_0(g_0)$, adapting $O(G \mid H, S_b)$ to H according to g_0. Now assume a change from environment H to H' such that

$$\mu(M_0(g_0) \mid H, K_0, S_b) > \mu(M_0(g_0) \mid H', K_0, S_b).$$

Given $0 < \delta < 1$, if $\mu(M_0(g_0) \mid H', K_0, S_b) \geq \delta$ then it is decidable if there exists $M_1(g_1)$ as an evolution $(M_0(g_0) \Rightarrow M_1(g_1))$ of $M_0(g_0)$ that will be better adapted to H' than $M_0(g_0)$.

Proof: There are two conditions to be considered:

1) If $\mu(M_0(g_0) \mid H', K_0, S_b) < \delta$ then $M_0(g_0)$ must be rejected and forbidden to evolve in H', and
2) If $\mu(M_0(g_0) \mid H', K_0, S_b) \geq \delta$, and since $\mu(M_0(g_1) \mid H', K_0, S_b)$ is dependent on both $\rho(d(s_0, s_t))$ and $\tau(s_t)$, then it is possible to evaluate:

a) if there exists any s_t whose relevance may be increased in order to augment $\mu(M_0(g) \mid H', S_b)$ such that

$$\mu(M_0(g_1) \mid H', K_0, S_b) > \mu(M_0(g) \mid H, K_0, S_b).$$

If this is the case, them a new goal g_1 is accepted to be a modification $(g_0 \Rightarrow g_1)$ of g_0. The condition to accept $g_0 \Rightarrow g_1$ is that g_1 continues to fulfill N. This is a decision of agents N based on the actual values of $\rho(d(s_t, s_n))$, because if $q(s_n) > 0$ then it is possible to increase $\tau(s_t)$ of s_t involved in $d(s_t, s_n)$;

b) if there exists any s_t whose $\rho(d(s_0,s_t))$ may be increased by changing the processing of $M_0(g)$ in any step from definition 3.12d to h, to augment $\mu(M_0(g)|H', S_b)$ such that according to theorem 3.5 then: $\mu(M_0(g) \mid H', K_1, S_b) > \mu(M_0(g) \mid H, K_0, S_b)$;

c) if a new g_1 may be generated $(g_0 \Rightarrow g_1)$ from g_0 by exchanging some $s_t \in V_g$ by some other new $s_{t'} \in V_g$ such that

$$\mu(M_0(g_1) \mid H', K_0, S_b) > \mu(M_0(g) \mid H, K_0, S_b) \text{ because}$$
$$\rho(d(s_0, s_{t'})) \circ \tau(s_{t'}) > \rho(d(s_0, s_t)) \circ \tau(s_t)$$

Those $s_t \in V_g$ selected to be exchanged must be those associated to $\rho(d(s_t, s_n)) \to 1$, because if the contents of g_0 are modified, then steps in definitions 3.12d – k and in definition 3.11 must be processed again.

If at least one of the above strategies succeed in promoting

$$\mu(M_0(g_1) \mid H', K_0, S_b) > \mu(M_0(g_0) \mid H, K_0, S_b) \text{ or}$$
$$\mu(M_0(g_1)|H', K_1, S_b) > \mu(M_0(g_0)|H, K_0, S_b)$$

then $M_0(g)$ will evolve into $M_1(g)$; the process of this transformation is called here evolutionary learning.

Corollary 3.6: Let there be a model $M_0(g_0)$ adapting $O(G \mid H, S_b)$ to H according to g_0. Let also g_1 be a transformation $(g_0 \Rightarrow g_1)$ of g_0 because the relevance of $s_t \in V_g$ is increased and/or $s_t \in V_g$ is changed.

If g_1 promotes

$$\underset{g_1}{\zeta} \rho(d(s_t, s_n)) \circ \tau(s_t) > \underset{g_0}{\zeta} \rho(d(s_t, s_n)) \circ \tau(s_t), \zeta \text{ as an S-norm}$$

then $M_1(g_1)$ is an optimization of $M_0(g_0)$.
Proof: From definition 3.11 and theorem 3.6.
Remark 3.13: Evolutionary learning has some distinctive properties:

a) First of all, it requires an initial knowledge K_o characterized by a set of m initial models to support the initial modeling $M_0(g_0)$ of H, guaranteeing the initial survival of $O(G \mid H, S_b))$ in H;
b) This K_0 is hardwired in the brain during embriogenesis, guided by phylogenetic information stored in $\{L(G(H, S_c))\}_{c=1 \text{ to } n}$;
c) The evolution of this K_0 is guaranteed by the ambiguity of G, that is, by the fact that $\rho(d(s_i, s_j) \mid H)) \rightarrow 0.5$, because in such a condition, the restrictions imposed by a changing H can be easily modeled by modifying $M_0(g_0)$ according to the Theorems 3.4, 3.5 and 3.6;
d) It is a sequentially ordered process, such that what is learnable at step a is very dependent on the evolution of K_0 up to this moment $K_0 \Rightarrow .... \Rightarrow K_a$;
e) C the ability $L(M_j(g_j))$ to learn a new model $M_j(g_j)$ is directly related to the similarity of this new model and any $M_a(g_a)$ composing the actual knowledge K_a; this similarity is $\mu(g_j, g_a)$. Thus:

$$L(M_j(g_j)) = \max_{K=1}^{a} \mu(g_j, g_k)$$

f) the capacity of learning $L(M_j(g_j))$ is also dependent on the ambiguity of $L(G \mid H, S_b))$ that defines the plasticity of the neural enrolment in the tasks required to solve $M_j(g_j)$.
g) if $M_j(g_j)$ may be learned as a modification of $M_a(g_a)$, then it is said to be an evolution of this latter model. Thus:

$$M_{a+1}(g_{a+1}) \Rightarrow M_j(g_j)$$

h) the capacity of learning $L(M_j(g_j))$ of $M_0(g_j)$ is also dependent on the restrictions imposed by H on the resources required by any g_j, g_a.
i) On the one hand, if resources are scarce, then competition may be established between $M_j(g_j)$ and $M_0(g_a)$, such that either $M_a(g_a) \Rightarrow M_{a+1}(g_{a+1})$ does not occur, or $M_a(g_a)$ is blocked if not extinguished.
j) On the other hand, if resources are plentiful, then $M_a(g_a)$ and $M_j(g_j)$ may reinforce each other in a symbiotic process, in order to have better odds to survive in $O(G \mid H, S_b))$
k) a model $M_j(g_j)$ may be stored in a set of $\{O(G \mid H, S_b))^i\}_{i=1 \text{ to } h}$ living together such that it is learned by organism $O(G \mid H, S_b))^s$ from another organism $O(G \mid H, S_b))^t$ that has previously learned or discovered $M_a(g_a)$.
l) In this context, discovering $M_j(g_j)$ is to learn it by itself, whereas learning $M_j(g_j)$ means to learn it from some other $M_a(g_a)$.

3.8 Summary

The brain $O(G(H, S_b))$ is defined herein as a DIPS in charge of processing a distributed language $\{L(G(H, S_{c,d}))\}_{d=1 \text{ to mc}}\}_{c=1 \text{ to n}}$. Each type of neuron S_c composing $O(G(H, S_b))$ is specialized in processing a given $L(G(H, S_c))$ supported by G, and the brain may contain up to m_c of such neurons. The language given by $L(G(H, S_{b,s}))$ describes the synaptic transactions $d(s_t, s_o | G, H, S_{c,i}, S_{d,j})$ between any two neurons $S_{c,i}, S_{d,j}$ of $O(G(H, S_b))$. The types of neuron and their quantity are initially defined by the set of homeobox genes O of G during the embriogenesis of $O(G(H, S_b))$, but post-natal learning changes both the number n of the different types of neurons and the number m_c of such cells. This type of learning is called here *evolutionary learning* to distinguish it from the classical connectionist learning procedures, $\rho(d(s_t, s_o | G, H, S_{c,i}, S_{d,j}))$. DIPS reasoning $R(G(g)| H, K, S_b))$ is defined by the set of transactions $d(s_o, s_t)$ fulfilling a goal g associated with a detected resource need. The conditions for the evolution of $R(G(g)| H, S_b))$ are set by both the ambiguity of $<\Omega(G| H, S_b)>$ of $G(H, S_b)$ and by the control capability $\theta(L(C| H, S_b))$ of $L(C,| G, H, S_b)$.

4 Neural Computational Mechanisms Supporting Cognitive Processes

In the previous chapter we proposed that the brain is a Distributed Intelligent Processing System (**DIPS**), with specialized agents for the tasks of gathering information from the environment, manipulating the information according to established goals, and implementing actions to reach those goals. Each cell in the brain is conceived of as a specialized agent that expresses a subset of the brain's language **L(G)**.

The performance of cognitive functions requires both cooperation and competition between specialized agents. Cooperation allows for the performance of functions, composed of sub-functions, which are performed by different units. Competition is a characteristic of selective processes (e.g., "winner takes all" mechanisms) that can help to increase the efficiency of the whole system in coping with environmental challenges. Both cooperation and competition are limited by possible mismatches between the subsets of the language handled by the different processing units.

An evolutionary strategy to avoid computational breakdown is proposed here wherein the brain employs quantum computation. This strategy allows microstate entanglement of several spatially distributed processing units, thus providing a supplementary communication channel to overcome possible mismatches. This channel also provides instantaneous binding of the informational content being processed in such units (a property which has been regarded as an essential ingredient for a theory of consciousness; see Rocha et al., 2001).

4.1 Basic Concepts of Quantum Computation

Two of the most surprising properties of quantum systems are microstate *superposition* and *entanglement*. Superposition is the coexistence of different microstate values of the same particle at the same time. Superposed states are reduced to a single state by the act of measurement or by other kinds of interaction with the macro-environment, which are called *deco-*

herence. Entanglement is a strong microstate correlation between spatially separated particles. It is an experimental finding still not explained in terms of causal processes based on electromagnetic or gravitational forces (the forces that operate at the macroscopic level). One interpretation of the phenomenon is that entangled particles behave as a single entity, despite their distributed spatial locations.

Quantum computation is a research area devoted to experimentally manipulating the superposition and entanglement of microstates, and to developing algorithms that could be implemented in such quantum systems. Of course, a superposition of states cannot be directly manipulated, since any interaction of the quantum system with the experimental apparatus reduces superposed states to a single one. However, the *combination* of superposition and entanglement can be manipulated through clever devices. While entangled, two or more particles have correlated, superposed states (e.g., spin up and down). Therefore, by measuring one particle, and thereby reducing its microstate to a single one (e.g., spin up), we gain knowledge about the actual microstate of its spatially separated, entangled partner(s) (e.g., spin down).

The above situation is equivalent to obtaining two bits from one. Quantum computation uses the concept of a quantum bit (*qubit*), which is equivalent to a superposition of orthogonal states in Hilbert space; i.e., **n** qubits corresponds to 2^n superposed states. Entanglement of two particles generates two qubits, corresponding to four classes of possible states, which are called *Bell states* (Zeilinger, 1998; see eq. 3 below).

A qubit can be in any state:

$$\alpha |1> + \beta | 0> \tag{4.1}$$

where α and β are complex numbers called amplitudes, subject to:

$$|\alpha|^2 + |\beta|^2 = 1 \tag{4.2}$$

Measurement on the system $\alpha | 1 > + \beta | 0 >$ results in the qubit making a probabilistic decision (Brassard, 1997; Brassard et al., 1998): with probability $|\alpha|^2$, the qubit takes the value $| 0 >$ and with complementary probability $|\beta|^2$, it equals $| 1 >$.

Because a physical system of n qubits requires 2^n complex numbers to describe its state, two qubits can be in the Bell states:

$$\alpha|00> + \beta|01> + \gamma|10> + \delta|11> \qquad (4.3)$$

such that

$$|\alpha|^2 + |\beta|^2 + |\gamma|^2 + |\delta|^2 = 1 \qquad (4.4)$$

Let a quantum computer **QC** be defined by Eqs. 4.3 and 4.4. A given instruction i_i may be written on it by, e.g., changing its state to

$$\alpha|00> + \beta|01> + \gamma|10> - \delta|11> \qquad (4.5)$$

that is, by modifying the condition of qubit $|11>$.

The change influences the entanglement of this qubit with its partner(s). Quantum computation is based on such an interference of qubits, which has been proved capable of performing like classical logical gates as well as other, so-called *quantum gates*.

The only constraint on quantum gates is *unitarity*, which is based on the constraint 4.2. Therefore, "any unitary matrix specifies a valid quantum gate!" (Nielsen and Chuang, 2000,p. 18). An important operation is instantiated in the Hadamard gate, whereby $|0>$ is changed into $(|0> + |1>)/\sqrt{2}$ and $|1>$ is changed into $(|0> - |1>)/\sqrt{2}$. Applying the Hadamard gate twice to a qubit generates an output that is identical to its initial state.

According to Nielsen and Chuang (2000), a QC has five requirements:

a) two parts, one classical and one quantum; although the classical part is not necessary, in practice it is useful in order to specify in binary logic the inputs and outputs to and from the quantum part;

b) a suitable state space; for n qubits the corresponding state space is a 2^n-dimensional complex Hilbert space;

c) the ability to prepare states in a computational basis (as pure entangled states);

d) the ability to act as quantum gates, preferably universal gates such as the Hadamard and Controlled NOT (CNOT; an equivalent of the classical XOR, or exclusive disjunction) gates; and

e) the ability to perform measurements in a computational basis.

Therefore, any proposal of biological quantum computation should be able to demonstrate the possibility of a biological structure to fulfill such requirements.

4.2 Cellular Processing

The cell is herein conceived of as a distributed stp system. The set of all possible stp strings processed by a given cell S_i is (see Definition 2.7):

$$Ł(G \mid H, S_i) = \{ d(s_0,s_t) \mid \rho(d(s_0,s_t) \mid H, S_i) \to 1 \} \quad (4.6)$$

Any cellular function $F(S_i)$ is therefore described by means of a set of coupled sentences $d(s_0,s_t)$ of $Ł(G \mid H, S_i)$, in the same way that a text is composed of a set of phrases. Thus,

$$F(S_i) = Ł(G \mid H, S_i) \quad (4.7)$$

The $F(S_i)$ biochemical process is catalytic, dependent on the action of enzymes. Any enzyme is a protein having two or more active sites able to promote interactions between other molecular elements, interactions that would not spontaneously occur, or would occurs at very low rates, in the absence of such proteins. A theory of cellular catalytic processes was proposed by Monod, Changeux and Jacob (1963). Called the theory of *allosteric* mechanisms, it states that there are two possible configurations for the active sites of a protein, called **T** (tense) and **R** (relaxed). The R state has higher affinity with the substrate than the T state. The "all-or-none" allosteric rule says that all sites are in conformation T or in conformation R (a case of exclusive disjunction; see Babloyantz, 1986). Therefore, if an effector acts upon an active site of a protein, changing it from T to R, then all other sites will automatically change to R (and vice-versa regarding changes from R to T).

The simplest case of an allosteric mechanism consists of a catalyst having only two active sites. In this case, binding of one site with an effector selects the configuration of both sites. If the first site changes from T to R, the second site also changes to R, whereupon it may become able to bind to a substrate and so change some property of that substrate. In sequence, the substrate may act as an effector by binding to a second catalyst, and so on, generating stp sentences and texts (for an update on allosteric mechanisms in the brain, see Changeux and Dahene, 2000).

There are two, more complex, cases of allosteric mechanisms, which will be mostly important for the formation of stp sentences and texts in the brain. These are known as: (1) **coincidence-detectors**: proteins whose spatial configurations are determined by a set of other molecules that have to bind it within a defined time interval after the first binding, as in the case of the membrane NMDA receptor and the cytosolic adenylyl cyclase; and also (2) **second-order catalysts:** proteins that control the phosphorylation—i.e., activation of sites—of the other proteins, illustrated by the family of kinases.

Coincidence detectors are proteins that have 3 or more active sites, and require *the conjoint activation of at least two sites* to generate a conformational change (transition from T to R, or vice versa in all sites).

Therefore, it is necessary that the binding of two or more effectors to two or more sites of the protein, in a fixed time window (depending on the properties of the protein), to generate the conformational change that, in turn, will promote its binding to the substrate, activating a biological function (i.e., triggering the formation of new words, sentences and texts).

In the above sense, coincidence-detection (**CD**) is a kind of self-organizing mechanism, by which states previously achieved in the system control the emergence of a new state. As the new state is necessarily correlated to the previously obtained ones, CD *increases the coherence* of the system as a whole, while at the same time promoting an increase in diversity. In other words, CD is a means whereby biological systems increase their organization while also decreasing entropy. (the alternative would be to increase organization by increasing redundancy, but in this case the strategy would fail, since beyond a given limit entropy would also increase and then the organization would decrease—a bell-shaped curve).

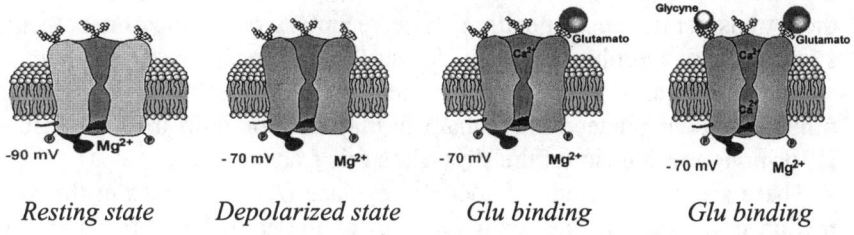

Resting state   *Depolarized state*   *Glu binding*   *Glu binding*

Figure 4.1 The NMDA channel: *a complex molecule that operates as a coincidence detecting device*

CD can be exerted on signals arriving at the cell and on internal stps at the cytoplasm. We exemplify the first kind with the NMDA receptor and the second one with the protein adenylyl cyclase. The NMDA receptor is a protein with several active sites, which binds to glutamate (Glu), glycyne (Gly), zinc (Zn), etc., and to one effector site (controlling its ion channel) that binds to magnesium (Mg; see Fig. 4.1). In the resting state, the protein adopts the R conformation, the one in which the effector site is active and bound to Mg. In this case, the ion channel inside the protein is blocked by Mg. This situation changes when the membrane is depolarized by activation of another channel (e.g., the AMPA channel), and Glu binds to its site. When these two events occur inside a temporal window (around 100 milliseconds), the protein changes from configuration R to T, and then the Mg that was bound to the ion channel site (and blocking it) is removed, allowing the entrance of Ca^{2+} ions from the extracellular milieu (see Bliss and Collingridge, 1993; Konig et al., 1996). Ca^{2+} entry activates several new stps inside the cell.

Definition 4.1: A CD stp is a sentence $d([s_1,...,s_c],s_t)$ having a complex $s_o=[s_1,...,s_c]$, and submitted to a temporal restriction Δ such that

$$\rho(d(s_o, s_t) \mid H, S_i, \Delta) \to 1,$$

if and only if the initial chemical transactions $\delta s_1 \gamma \to ... \to \delta s_c \gamma$ occurs within the time window Δ (see Definition 2.3).

Changeux and Dahene (2000) have argued that other kinds of ligand-gated channels, such as the nicotinic acetylcholine receptor, can display the properties of a temporal coincidence detector: "a large majority of ligand-gated ion channels display desensitization and/or potentiation with kinetics which may be fitted by allosteric models. In addition, because of their transmembrane disposition these receptors carry sites on both their synaptic and cytoplasmic sides, letting the molecule integrate within a given time window multiple convergent pre-and post-synaptic signals. A time coincidence detection mechanism may then be built from the discrete all-or-none mechanism of the slow allosteric transitions".

There are eleven kinds of adenylyl cyclase (AC) enzymes in the mammalian brain, having the main function of generating cAMP from ATP. They have been thought to be molecular coincidence-detectors (Anholt, 1994) also, since they link metabotropic receptors as well as G-proteins to cAMP signaling pathways and also to Ca^{2+}-activated calmodulin, which are essential for many brain functions. Therefore, AC can be conceived of as a coincidence detector for intracellular processes that occur in a longer time window than obtains for the NMDA receptor. While the latter is adequate for perceptual processes, AC could be better suited to the timing of mnemonic and emotional processing.

Both Ca^{2+} and G-protein pathways lead to the activation of AC, cAMP release and then to kinase (second order catalyst) activation. Therefore, the activity of the kinase family of enzymes is a part of all stps activated by transmitter-receptor binding at the neuronal membrane. A diagram of some stps activated by glutamate receptors, including two kinases (PKA and CamKII), and their possible role supporting conscious processing is shown in Figure 4.2.

The kinase enzyme family has a major function in cell metabolism: the catalysis of phosphate transfer from ATP to a protein substrate. Through the action of kinases, other proteins becomes activated and can perform a variety of biological functions. Therefore, kinases are second-order catalysts, the catalysts that control the other catalysts.

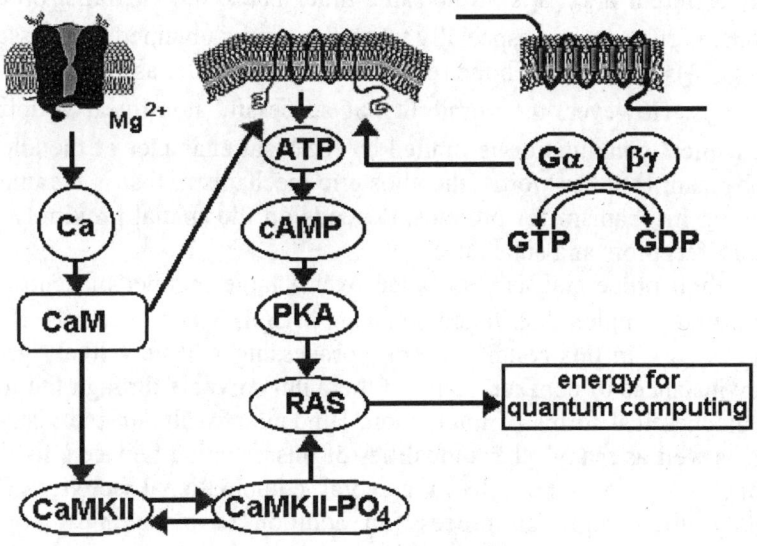

Fig. 4.2. An example of $d([s_1,...,s_c],s_t)$: *providing the energy for quantum computing*

A recent genome mapping (Kostich et al, 2002) revealed that the kinase family of enzymes corresponds to 510 sequences in the human genome, approximately 2% of the total genome. The kinases determine several operating characteristics of neurons by controlling the proteins that control a variety of functions, as cytoskeleton spine density, post entry Ca^{2+} signaling, activation of transcriptional factors (CREB), and control of glucose metabolism. They also control the life (cell growth and division) and death (apoptosis) of neurons. The common aspect of all these functions is the transfer a group of atoms (the phosphoryl group) between different molecules. This mechanism, that is proper to the kinases, has been compared to a switch that causes biochemical pathways to work slower or faster.

The role of some members of the kinase family in the brain, in learning and memory processes, has been well studied by several researchers (see Sweatt, 1999, for a review). Mainly protein kinase A (PKA), protein kinase C (PKC), calmodulin-dependent protein kinase II (CaMKII), and mitogen-activated protein kinase (MAPK) have been implicated in such processes.

It is clear from Fig. 4.2 that the processing of $Ł(G \mid H, S_i)$ by the cell S_i results from a complex set of concurrent molecular transactions involving many different $d(s_O, s_t)$s at the same time. These huge parallel processes become very complex, especially when the results obtained in an assembly of $d(s_O, s_t)$s doesn't match the results obtained by other assemblies of other $d(s_O, s_t)$s. However, the possibility of successful non-linear coupling in biochemical computation is limited by the local character of the allosteric mechanism. In other words, the allosteric mechanism, that is the universal coupling mechanism for proteins, depends on the spatial proximity of effectors, receptors and substrates.

Cerebral processing is distributed over a huge number of neurons $S_{i,n}$, supporting complex distributed language $Ł(G(H, S_b)) = \{\{Ł(G(H, S_{i,n}))\}_{n=1 \text{ to } k}\}_{i=1 \text{ to } l}$. In this condition, brain processing will very likely generate many instances of conflict, many of them not solvable through the foregoing biochemical forms of interaction. Binocular rivalry in the visual system, as well as many other modalities of mismatching between distributed parallel units, are the rule in cerebral computational activity. Consequently, the integration process, in addition to biochemical coupling, should involve non-local interactions between assemblies that recognize different features of the stimulus and/or represent different goals to be achieved through voluntary action. It thus ought to involve quantum computations. In this context, CD stps will help to promote superposition and entanglement.

4.3 Current Physical Implementations of Quantum Computers

Although most experimental results obtained with quantum computing are limited to a few particles encoding a small (e.g. 2 to 4) number of qubits, the properties of quantum computing are perfectly suitable for the modeling of brain function. The level of integration of millions of parallel processing units in the brain cannot be explained through a serial coupling of operations, which would lead to an extremely long processing chain, or by convergence to a single region, which would be overloaded. The brain surely has some convergence regions, but this architecture is limited to regional integration and cannot explain how the results of millions of processors are bound together in the performance of a complex cognitive task, or in the generation of conscious states.

If brain function resembles the operations of a QC, the most relevant property to be highlighted is *entanglement*. This property allows any number of particles, belonging to millions of cells, to instantaneously bind their internal quantum informational states, generating a unified waveform that would correspond to a moment (or to a temporal unit) of consciousness. Also, quantum cryptography may be required to explain how one single sensory (e.g. visual, auditory, etc.) quantum- computational line of processing would not be corrupted by simultaneous computations carried on in each of these sensory modalities (this subject will be discussed in Chap. 5).

Nuclear Magnetic Resonance (NMR) is being successfully used to built QCs (Warren, 1997). Sequences of radio frequency pulses (**rfp**) manipulating spin orientations constitute quantum logic gates and perform unitary transformations on the QC's state. In the first stage rfps create state superposition. In the sequence, rfps modify spin orientations to write the desired instruction. Finally, rfps change state amplitudes to enhance the probability of the desired answer. NMRQCs have been built based on ^{1}H and ^{13}C nuclei in chloroform, on ^{3}Ce atom (Ahn et al., 2000), using two ^{1}H nuclei in cytosine, as well as on alanine and trichloroethylene. Since quantum computing may be implemented over organic molecules, in particular on amino acids, it will be assumed here that state superposition may be achieved on specific proteins.

Another technique being proposed to build QCs employs ions in microtraps (Cirac and Zoller, 2000). In this approach, the QC is composed of a set of **N** ions confined in independent harmonic potential wells, which are separated by a constant distance **d**, large enough to prevent Coulomb repulsion from exciting the vibrational states of the ions, and so allows the ions to be individually addressed (Fig. 4.3).

Quantum information is encoded in the *internal states* of particles (as the spins of electrons or protons, and electronic states of ions), but the transfer and retrieval of such information requires the manipulation of *external* states (values related to movement). Therefore, quantum- informational devices are *two-level systems* (Nagerl et al., 2000). Experimental manipulation of such systems is based on the achievement of a strong coupling between external and internal states. In ion-trap computers, internal states of one ion are coupled to the vibrational degrees of freedom of the ions in the trap (Nagerl et al., 2000) in such a way that the emission or absorption of phonons (quasi-particles) by the ions, caused by a change in movement, is accompanied by changes in internal states.

In the ion-trap quantum computer, the coupling between the vibrational and the internal states of the ions is made by an oscillator. Laser pulses have been used as oscillators, allowing the transferal and retrieval of information to the internal states of ions. Rfps forming a quadripole control the movement of a population of ions in a magnetic trap. Laser beams are used to individually access the ions and change/measure their internal states (Kielpinski et al., 2002).

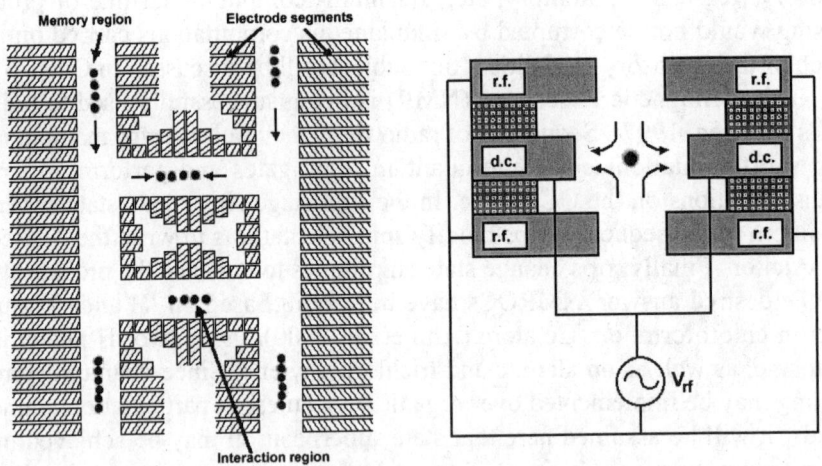

Fig. 4.3. Ion-Trap QC *(adapted from Kielpinski et al., 2002)*

In this setting, the CNOT logical gate has been implemented with a degree of realibility around 75% (Schmidt-Kaler et al., 2003). A target ion has its internal state changed to the state labeled as 1 in binary code. The controlled ion is then changed (from 1 to 0 or from 0 to 1) only when the state of the target ion is 1. As theoretically predicted, the experiments have shown that once the string of ions in the trap gets entangled, the change in the internal state of the target ion naturally produces the expected change in the controlled ion(s).

4.4 The Dendritic Spine as Quantum Computing Device

There are several kinds of chemical elements and molecules that together with their constituent particles could be used by the brain to generate quantum computational systems. Among the experimentally proven alternatives reviewed in the last section, we have chosen trapped Ca^{2+} ions as a plausible neurobiological model, by considering the already known functions of such cations, which are frequently trapped in the dendritic spine or in intracellular compartments (for a review of Ca^{2+} functions see Alkon et al., 1998; Ghosh and Greenberg, 1995). These Ca^{2+} movements are central to all kinds of cognitive and emotional processes in the brain. Ca^{2+} ions trapped in cellular compartments would quite arguably satisfy the requirements for a quantum computing device, while retaining information from the biochemical processes they participate in.

Spines are morphological specializations of mammalian neurons that receive the majority of synaptic excitatory inputs. Their role has recently been studied by using new techniques, such as two-photon microscopy and the generation of transgenic mice. A consensus has emerged among neuroscientists that "spines are biochemical units that compartmentalize calcium...the specific function of spines could be calcium-derived signal transduction, restricted to individual synaptic inputs, which could implement input-specific learning rules" (Holthoff et al., 2002).

By imaging spines from layer V pyramidal neurons of the rat primary visual cortex, the above authors discovered that the position of the spine in the dendrite is related to its Ca^{2+} dynamics and to its putative visual processing rules.

Spines are suited to the accumulation of Ca^{2+} since they have a "head" that is larger than the "neck" that connects it to the dendrite (see Sabatini et al., 2001, 2002). The STPs internal to the spine are compartmentalized by several scaffold, anchoring and adaptor proteins (see Fig. 4.3).

Ca^{2+} enters into the spine through AMPA and NMDA membrane receptors, and also through voltage-dependent calcium channels (VDCCs). Such membrane receptors are ion channels controlled by the binding of a ligand (mainly the neurotransmitter Glu). VDCCs, in turn, are ion channels controlled by the voltage (action potential) of the membrane. The three kinds of Ca^{2+} channels work cooperatively. AMPA and VDCCs can work independently of the other channels, while the NMDA channel is dependent on the others, because of its complex physiology. Being an activity-dependent mechanism, NMDA opening obeys a more restrictive set of conditions. On the other hand, it works as the faster controlling mechanism of massive Ca^{2+} entry, and therefore has a central role in the processing of afferent information.

A central factor determining Ca^{2+} dynamics in the spine is the presence of Ca^{2+} pumps, which cause its differential distribution among the spine's internal compartments. Holthoff et al. (2002) also found that the Ca^{2+} pump's strength is positively correlated with its distance from the soma of the neuron. Ca^{2+} ions trapped in the dendritic spine are moved from one internal compartment to another, or through the neck to the endoplasmatic reticulum, by the action of Ca^{2+} pumps.

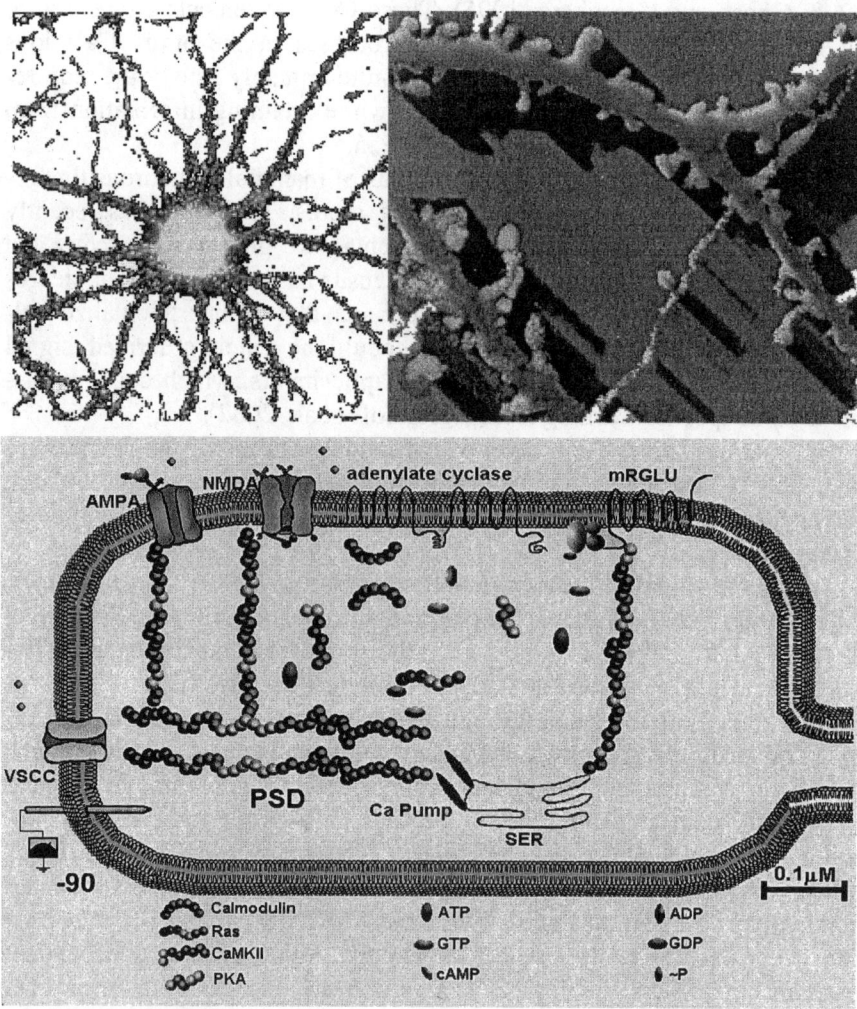

Fig. 4.3. **The dendritic spine**: the dendritic spine has a very complex structure supporting advanced functions

The energy prompting Ca^{2+} movement in the spine is made available through the reduction of ATP by one member of the kinase family, the calmodulin-dependent protein kinase (CaMKII). This kinase is capable of autophosphorylation, or self-catalysis. It can also be activated by binding to calmodulin (CaM), a protein that is activated by Ca^{2+} and available in a large number of configurations, which are selected by the entering Ca^{2+} (see discussion in Pereira Jr., 2003).

The cognitive profile of CaMKII is completely adequate to the new, putative functions we assign to it. CaMKII has been proposed as a candidate for the molecular basis of memory: "The analysis of CaMKII autophosphorylation and dephosphorylation indicates that this kinase could serve as a molecular switch that is capable of long-term memory storage" (Lisman et al., 2002). Also, Vianna et al. (2000) suggest that "memory formation of spatial habituation depends on the functional integrity of NMDA and AMPA/kainate receptors and CaMKII activity in the CA1 region of the hippocampus [and that] the detection of spatial novelty is accompanied by the activation of at least three different hippocampal protein kinase signaling cascades".

The CaMKII cycle in the dendritic spine has recently been elucidated (Fox, 2003). Its movement in the spine is controlled by Ca^{2+}-induced excitatory activity. In turn, CaMKII provides the energy for Ca^{2+} movements in cellular compartments. These movements, we hypothesize, would be related to the encoding of information in internal electronic states of the ions. Considering also the existence of surrounding fields generated by the movements of other ions (Na+, K+), the dendritic spine has an impressive similarity with current ion-trap quantum computers, which use the same Ca^{2+} caged in magnetic traps, moved by magnetic quadripoles and coded/decoded by laser beams.

In our framework, entanglement of Ca^{2+} internal states in spines would be generated by the coordination of actions of membrane channels and cellular agents that manipulate this cation. For instance, the NMDA channel CD would transform a temporal coincidence of Glu pulses generated by perceptual mechanisms into a quantum coherent state by erasing the temporal order and altering the Ca^{2+} vibrational state to generate superposition (see the conditions for the generation of entanglement in Bouwmeester and Zeilinger, 2000).

The possibility of the dendritic spine working as a Ca^{2+} trap QC depends on the demonstration of its compatibility with the requirements for a physical system to perform quantum computation. Since the spine has mesoscopic mechanisms that manipulate individual atoms, we cannot rule out the possibility that such mechanisms could prepare the ions in superposed and entangled states, and then perform transformations corresponding to logical gates by means of metabolic operations on the trapped Ca^{2+}s. The conjoint activation of AMPA, NMDA, and VDCC would control different quantum computational operations in spines. Neurobiological data has shown (see Krystal et al., 1999) that the balance of Ca^{2+} channels is necessary for the generation of normal consciousness, while disturbances (e.g., AMPA and VDCCs activation without NMDA) generate perceptual distortions and hallucinations (see Pereira Jr. and Johnson, 2003). These phenomena would be explained by the specific role of each channel in spinal QC, as proposed in more detail in the following chapter.

A popular objection to neural quantum computation derives from the warm and noisy conditions of the brain, as compared to the experimental settings where QCs have been realized (see discussion in Pereira Jr., 2003). However, it should be noted that Ca^{2+} ions are cooled in artificial ion-trap QCs to improve the accessibility of initial and output states as classical binary states (an important requirement for digital computers, corresponding to Nielsen and Chuang's last requirement), not for the generation and manipulation of superposition/entanglement. Isolation is important to assure the unitarity (also referred as the "reversibility") of the operation in the time window necessary for the artificial device to encode and decode binary data reliably. Such experimental constraints have not been proved to forbid natural quantum computing from occurring in shorter time windows or with less reliability than artificial QCs. Isolation and low temperatures are not therefore absolute requirements for the existence of QCs, but only conditions for the engineering of artificial QCs as digital computers.

5 The Brain and Quantum Computation

Quantum computing is becoming a reality both from the theoretical point of view as well as from the realm of practical applications. New techniques are being discussed and implemented, making quantum operations feasible at room temperature. Moreover, new architectures are being proposed for future quantum machines. Nuclear Magnetic Resonance and Ion Traps (IT) are among the technologies most used in quantum computation experiments.

The brain is the most sophisticated processing machine developed by nature thus far. Quantum information and quantum computation have been considered important issues in the effort to understand neural function and the phenomenon of consciousness.

Dendritic spines (DS) are specialized synaptic structures with low endogenous Ca^{2+} buffer capacity, allowing large and extremely rapid Ca^{2+} changes under physiological conditions. Ca^{2+} diffusion to the dendrite across the spine neck is negligible, and the spine head functions as a separate compartment on long time scales, allowing localized Ca^{2+} build-up during trains of synaptic stimuli. Also, DS are very plastic structures, involved in both rapid learning, or *imprinting*, as well as in the slower learning due to environmental changes.

Here DS are assumed to be IT devices that may be used to build Quantum Charge-Coupled Computers with an architecture similar to that proposed by Kielpinski et al (2002). We use the Deutsch-Josza algorithm (see a review in Nielsen and Chuang, 2000) to propose a model of a Quantum Cortical Pattern Recognition Device (QCPRD). Since the Deutsch-Josza algorithm is also probabilistically computable in a constant number of steps, we implemented a probabilistic QCPRD version and tested it as a Visual Pattern Recognition Device (VPRD).

We propose that quantum processing controls the allosteric **(R,T)** states of proteins which are involved in coincidence detector stps $(d([s_1,...,s_c],s_t))$. In this condition, superposition, entanglement and quantum processing are assumed to be dependent on phosphorylation of specific kinases, triggered by the entry of calcium through the NMDA channels and mR-Glu in the support of higher brain functions such as consciousness and working memory.

5.1 The Dendritic Spine as an Ion Trap Quantum Computing Device

An ITQC involves:

1. confining ions in a single narrow trap;
2. employing ionic electronic and motional states as qubit logic levels; and
3. transferring quantum information between ions through their mutual Coulomb interactions.

Ion trapping (Fig. 5.1) is realized by using a combination of radio frequency and quasi-static electric fields generated by a set of electrodes (Kielpinski et al, 2002). By varying the voltage on these electrodes, ions are confined in a particular region or transported along the local trap axis. Applying radio frequency voltage to outer layers creates a quadripole field that confines the ions transverse to the local trap axis (see Fig. 4.3). Laser pulses are used to implement quantum gates (see Kielpinski et al, 2002). Photodetectors are used to read the results of the quantum computation.

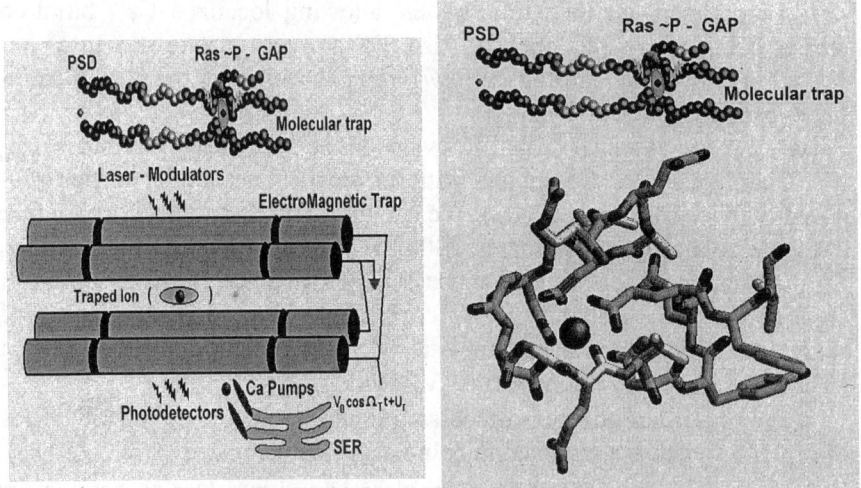

Fig. 5.1. The dendritic spine as a Quantum Computing Device: nature has endowed the brain with a huge computational capacity.

5.1 The Dendritic Spine as an Ion Trap Quantum Computing Device

We propose here that Ca^{2+} ions are trapped in molecular cavities of some special molecules composing the Post-Synaptic Density (PSD) and that energy released by phosphorylation of other PSD proteins implement different kinds of quantum gates. At the end of the computation, Ca^{2+} ions are moved into the spine's endoplasmatic reticulum (SER) or removed toward the extracellular space, depending of the obtained results.

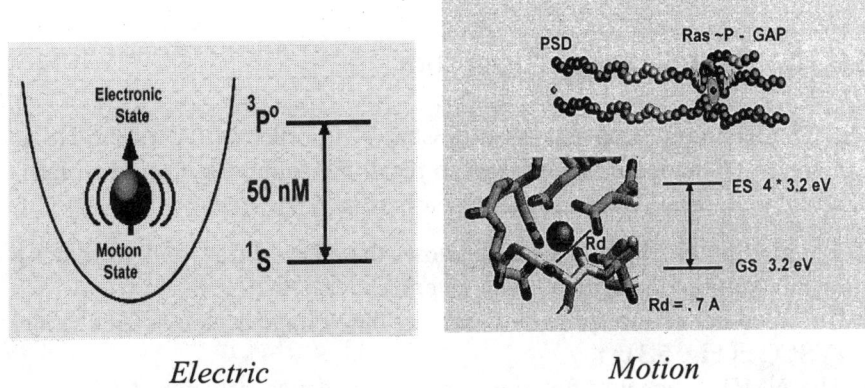

Electric   *Motion*

Fig. 5.2. The DS qubits: *building the neural qubits*

The electronic qubits for Ca^{2+} are[1]:

$$|0> = s^1 \text{ and } |1> = {}^3p^0 \qquad (5.1)$$

and the energy E required to promote the state transition (Harris, 1996)

$$s^1 \rightarrow {}^3p^0 \qquad (5.2)$$

is equal to 25 .21507 eV such that the frequency v of the activating photon is calculated from (Harris, 1996):

$$E = h\nu \qquad (5.3)$$

where h is Planck's constant. Its value is 50 nm (UV). The motion qubits are

$$|0> = \text{ground state; and } |1> = \text{first excited state} \qquad (5.4)$$

with ground state energy of 3.2 eV calculated for a cavity of around 0.7A. This value was chosen because Ca^{2+} radii is smaller than the atom having

---

[1] http://physics.nist.gov/cgi-bin/AtData/levels_form

the same number of electrons in the outside shell, which corresponds to Argonium.

Remark 5.1. It is interesting to note here that the frequency of energy absorption for Tyrosine transition electronic state is around 250 to 269 nm. Tyrosine is one of the aminoacids composing a family of kinases, called Tyrosine Kinases.

5.2 The Deutch-Josza Algorithm

The DS physiology was discussed in Chap 4. In order to understand DS as a quantum IT device it is sufficient to recall the following from the physiology of glutamate (Glu) and its receptors (Fig. 5.3):

1. Glu binding to AMPA receptors allows the entry of Ca^{2+} and promotes a depolarization of the membrane potential;
2. Ca^{2+} also enters the spine by means of the voltage sensitive Ca channel (VSCC in Fig. 5.3);
3. The NMDA receptor functions as a coincidence detector (CD), since depolarization removes the Mg^{2+} attached to it, and a posterior Glu-binding, within a temporal window Δ, promotes the entry of Ca^{2+}. Next, calcium-bonded Calmodulin (CaM) takes energy from ATP sources and delivers it to other biochemical processes;
4. Glu binding to metabotropic receptors (mR-Glu) promotes the activation of many types of G-proteins which control other processes;
5. Glu receptors are anchored in the membrane by a set of proteins; and
6. Ca^{2+} concentrations inside the cell are controlled by systems moving it to other organelles (e.g., SRE in Fig. 5.3).

Conjecture 5.1. The NMDA coincidence detector paradigm is proposed to create entangled quantum states in the brain, as follows:

1. A first Glu released by a pre-synaptic cell is received by the AMPA channel, promoting EM depolarization and Mg^{2+} release from the NMDA channel (Fig. 5.3A).
2. EM depolarization opens the VSSC channels; at this juncture Ca^{2+} enters and is trapped at the Post-Synaptic Density (PSD) (Figs. 5.1 and 5.3A).
3. A second release of Glu (by another pre-synaptic cell), within the temporal window Δ, transports Ca^{2+} to bind to CaM. CaM provides energy to create state superposition of other proteins, or trapped Ca^{2+} ions (Fig. 5.3B).

4. A third Glu release (by a third pre-synaptic cell) over GluMR activates G-proteins and then implements quantum gates with the entangled Ca^{2+} (Fig. 5.3C and D); and
5. Ca^{2+} moves into the SRE at the end of the quantum computations.

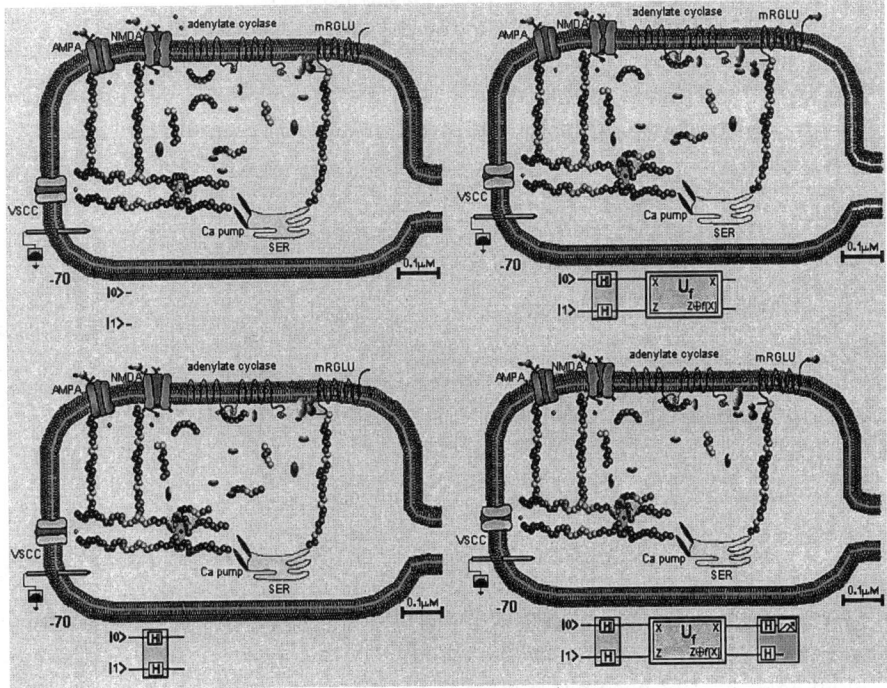

Fig. 5.3. The dendritic spine as a Quantum Computing Device: *using molecular machinery for quantum computation*

The Deutsch-Jozsa algorithm (DJA) was the first explicit example of a computational task performed exponentially faster using quantum effects instead of classical means (Nielsen and Chuang, 2000).

Given a one-bit function f, one has either two constant functions:
$$f(0) = 0 \text{ and } f(1) = 0, \text{ or } f(0) = 1 \text{ and } f(1) = 1;$$
or two "balanced" functions:
$$f(0) = 0 \text{ and } f(1) = 1, \text{ or } f(0) = 1 \text{ and } f(1) = 0.$$

A one-qubit QC (Fig. 1B) can decide whether f is constant or balanced in just one step.

DJA is proposed as a mechanism to provide a one-step answer to the following question: Is f constant or balanced?

Algorithm 5.1: DJA works as following (Fig. 5.3):
1. Starting with the standard state $|0\rangle\,|0\rangle$ at the input (**I**) and output (**O**) registers;
2. a NOT operation is applied to I;
3. Next, the Hadamard transformation (H in fig. 5.3) is applied to both registers.

Thus:

$$|0\rangle|0\rangle \xrightarrow{\text{NOT}} |0\rangle|1\rangle \xrightarrow{\text{H}} \left(\frac{|0\rangle+|1\rangle}{\sqrt{2}}\right)\left(\frac{|0\rangle-|1\rangle}{\sqrt{2}}\right) \quad (5.5)$$

In the sequence, the unitary transformation $\boldsymbol{U_f}$ is applied to both registers

$$\boldsymbol{U_f}\text{:}\left(\frac{|0\rangle+|1\rangle}{\sqrt{2}}\right)\left(\frac{|0\rangle-|1\rangle}{\sqrt{2}}\right) \rightarrow \left(\frac{1}{\sqrt{2}}\sum_{x\in B}(-1)^{f(x)}|x\rangle\right)\left(\frac{|0\rangle-|1\rangle}{\sqrt{2}}\right) \quad (5.6)$$

5. In this condition, O remains in the state $(|0\rangle - |1\rangle / \sqrt{2})$. If f is constant I is $\pm (|0\rangle - |1\rangle/\sqrt{2})$ and if f is "balanced" it is $\pm (|0\rangle + |1\rangle/\sqrt{2})$. If H is applied again to I, it becomes $\pm (|0\rangle)$ if f is constant and $\pm (|1\rangle)$ if f is "balanced".
6. Finally, these O states are reliably distinguished by a measurement in the standard basis, thus distinguishing balanced from constant functions after just one query.

Conjecture 5.2: DS process the DJA as follows (Fig. 5.3).

1. The AMPA channel is associated with the **O** register, whereas the VSSC is assigned to the **I** register;
2. The initial EM depolarization promoted by Ca^{2+} entry through the AMPA channel removes the Mg^{2+} from the NMDA channel and opens VSCCs;
3. Next, NMDA channels are activated, and CaM is used to perform the Hadamard transformation upon the Ca^{2+} trapped in the spine head;
4. Next, mR-Glu receptor is activated, and a G-protein is used to implement $\boldsymbol{U_f}$, as well as

5.2 The Deutch-Josza Algorithm

5. to perform another Hadamard transformation, and
6. the result is obtained by moving (or not) the Ca^{2+} into SRE.

Remark 5.2: A point worth remarking is that DS is amenable to many different and sophisticated experimental manipulations, which may be used to check the above assumptions.

Now, let's consider another function $f : B^n \rightarrow B$ that is either constant if the 2^n values are either 1 or 0, or balanced if exactly half (i.e. 2^{n-1}) of the values are 0 and half are 1.

Algorithm 5.2: DJA is extended to answer if $f : B^n \rightarrow B$ is constant or balanced, if

1. one starts with a row of n I qubits and one O qubit and to applies the same step procedures above. At the end, the n Is are in state
2. a NOT operation is applied to I;
3. Next, the Hadamard transformation ($H^{\oplus n}$) is applied to all n qubits of the I register and to the O register;
4. In the sequence, the unitary transformation U_f is applied to both registers, resulting in

$$|\xi_f\rangle = \left(\frac{1}{\sqrt{2}^n} \sum_{x \in 2^n} (-1)^{f(x)} |x\rangle \right) \quad (5.7)$$

5. If f is constant then ξ_f will be just an equal superposition for all the $|x\rangle$'s with an overall plus or minus sign, whereas if f is a balanced function then ξ_f will be an equally weighted superposition with exactly half of the $|x\rangle$'s having the minus signs;
6. Recalling that H has its own inverse (HH=1) and that H applied to each qubit $|0\rangle$ of $|\xi_f\rangle$ results in an equal superposition of all $|x\rangle$'s. Therefore, if f is constant then the resulting state is

$x = \pm |0\rangle|0\rangle...|0\rangle$, and if it is balanced, then $|x\rangle$'s is $x \neq |0\rangle|0\rangle...|0\rangle$.

Remark 5.3. The reading of each of the n qubits completes the measurement. DJA requires $O(n)$ steps to distinguish balanced from constant functions, whereas classical algorithm demand $O(2^n)$ steps for the same task. However, a probabilistic algorithm is able to solve the same task in k steps with a probability of $(1-\zeta)$ for correct answer and ζ less $1/2^k$.

5.3 The Quantum Cortical Recognition Device

The cortex is composed of a ordered set of layers (Fig. 5.4.a) numbered 1 to 6 from outside to inside. Their output neurons are called pyramidal cells. Input information from thalamus arrives at layer 4 to the stellate cells, which in turn conveys it to the pyramidal cells at layers 2 and 3. The output from these layers are distributed over other cortical areas, whereas the pyramidal cells from layer 6 send its axons to the thalamus, colliculus and brainstem, besides returning collateral branches to layer 2 and 3, and to layer 1. In this layer, the axons form the so-called parallel fibers also spreads to other cortical areas.

Many interneurons operate in the cortex, controlling the traffic of information between the stellate and pyramidal cells and between pyramidal cells. Many of these interneurons are GABA inhibitory neurons.

Many modulating circuits (RAS, 5-HR, DA and NA in Fig. 5.4.a) control the excitability of the pyramidal cells. These control circuits are in charge of setting the electrical code of the pyramidal cells as either a *tonic* or *phasic* encoding (Rocha, 1997), by depolarizing the membrane.

5.3.1 The Model

Definition 5.1: The electrical encoding by the pyramidal cell is:

1. in hyperpolarized states, when the membrane electrical potential EM approaching the K equillibrium potential E_k is supported by a phasic encoding (see Rocha, 1997 and Rocha et all, 2001) of short words formed by bursts of action potentials. The phasic encoding is mainly determined by an oscillating Ca^{2+} current. Also, hyperpolarized states favors the Mg^{2+} binding to the NMDA channel; and
2. in depolarized states, when the membrane electrical potential EM is growing less negative, supported by a tonic encoding of long words formed by a continuous spiking. The tonic encoding is mainly determined by Na^+, Ca^{2+} and K^+ ions. Also, depolarized states favor the Mg^{2+} uncoupling from the NMDA channel.

Thus if a neuron $n_i \in O(G \mid H, S_b)$ is able to express two languages

$$\mathcal{L}(G\mid H, n_{i,1}), \mathcal{L}(G\mid H, n_{i,2})$$

there must exist $d(s_0, s_c) \in \mathcal{L}(C, \mid G, H, S_b)$ to specify which of these languages are to be used in a given processing.

5.3 The Quantum Cortical Recognition Device

Model 5.1: Let the the DJA processing device CPRD in Fig. 5.4.b be formed by:

1. a retina R as an array of $r = p \times p$ bits;
2. a device B (cortical stelate cells) of m states of 1 bit;
3. a device Q (cortical pyramidal cells) of m states of 1qubit;
4. a device D (cortical stelate cell) of 1 qubit; and
5. a measuring device M of 1 bit.

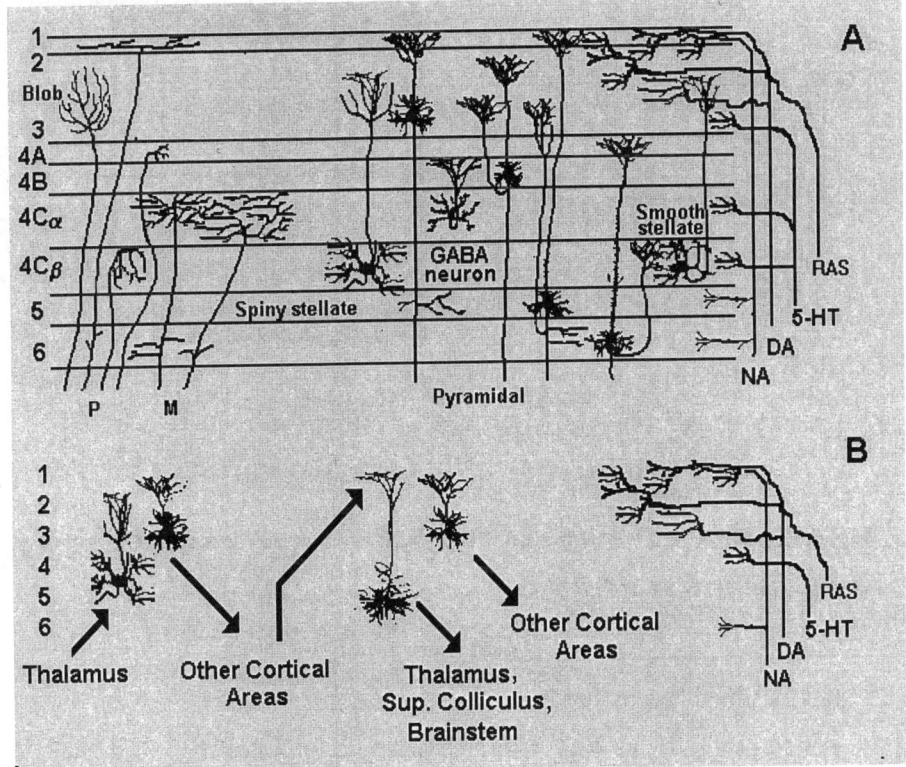

Fig. 5.4.a. Cortical layers and neurons (A) and connections (B)

A pattern recognition device is constructed by:
1. M_1 mapping neighbor n_1 (n_1=1 in Fig. 5.4) bits of R to a bit of Q;
2. M_2 mapping neighbor n_2 bits of R to a bit of B;
3. one to one mapping M_3 from B to Q;

4. one to all mapping M_4 from D to Q,
5. one to all mapping M_4 from Q to A;
6. a function $f(x)$, such that either $f(0) = f(1)$ and it is said constant or $f(0) \neq f(1)$ and $f(x)$ is said to be balanced.

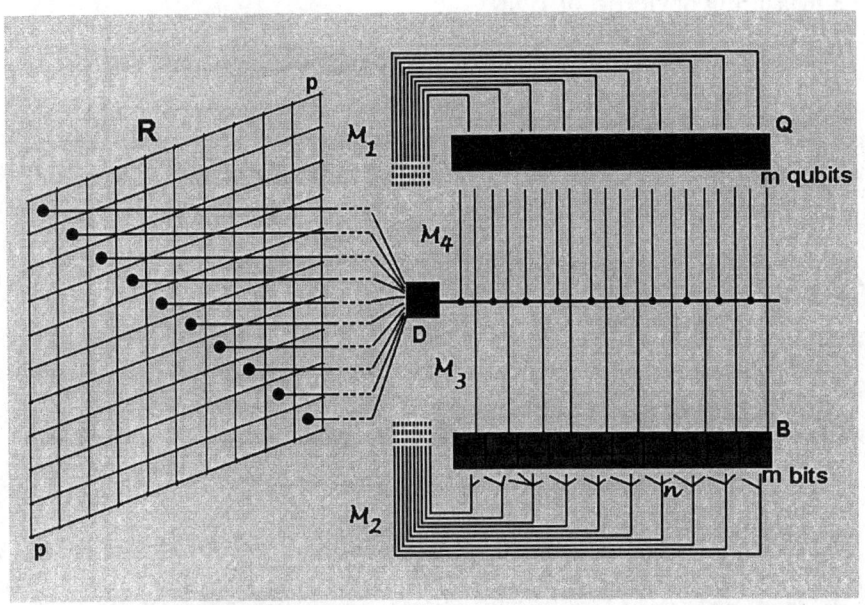

Fig. 5.4.b. The Cortex and the QCPRD: *building a powerful pattern recognition device using quantum computing*

It is assumed herein that:

1. stellate cells B in Fig. 5.4 are strongly inhibited by OFF stellate neighbors and excited by ON cells, such that
2. they clearly recognize and amplify edges (for contrast) in a given direction. Also:
3. pyramidal cells Q in Fig receive information from:

- Edge Detector Stellate Cells B;
- ON stellate cells activated by the retinal cells; and
- D stellate cells: on cells with large receptive fields.

5.3 The Quantum Cortical Recognition Device

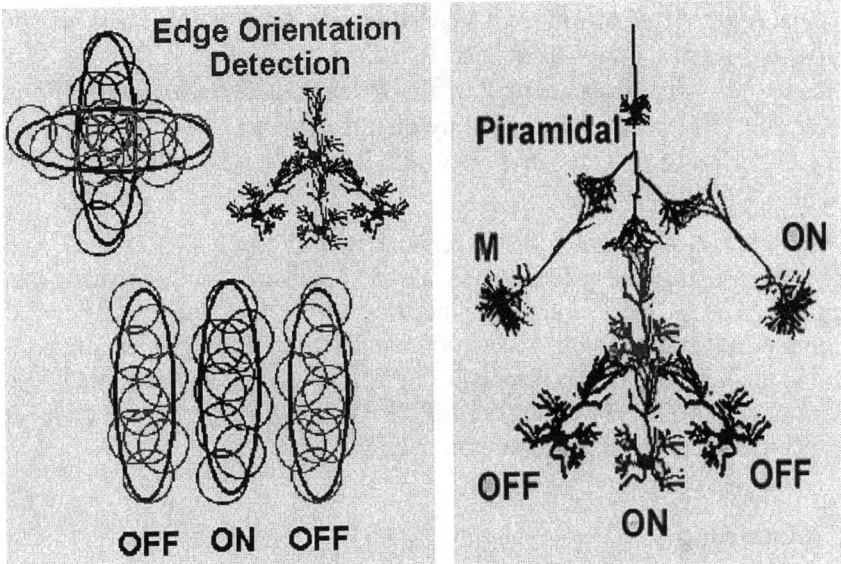

Fig. 5.5. The cellular structure of QCPRD: *enhancing contrast to detect borders.*

Algorithm 5.3: CPRD is assumed here a QCPRD if:

1. the D bit is set 1 at the moment t_1 if all m bits of R mapping to it are set 1 at time t_o ($t_o < t_1$) otherwise it is set 0;
2. the input register I (or VSCC channel) of Q is set by R and the output register (or AMPA channel) is set by D at time t_1;
3. If $0 < t_1 - t_o < \alpha$ then m bits of Q are set in superposition by Hadamard gates H triggered by the NMDA channel, and
4. a bit of B is set to 1 at the moment t_1 if the n neighbor bits of R mapping to it are set 1, otherwise it is set 0;
5. the function $f(x)$ at the ith bit of R is set at time ($t_1 < t_2$) by the mR-Glu system as constant if the ith bit of B is 1;
6. otherwise, the function is set as balanced, and the m qubits of R are transformed at the time t_3 by the unitary matrix

$$U_f : |x, d> \rightarrow |x, d \oplus > f(x)$$

7. after this, the IP_3 channel applies another H to I at time t_4;

8. M measures states of I in O(n) steps at time t_5, such that its bit is set 1 if $f(R)$ is constant, otherwise it is set 0, and
9. finally, the stimulus pattern P at the retina R is recognized if $f(R)$ is constant. This pattern is defined by the maps M_1 and M_2, and it is a diagonal in Fig. 5.4. The complexity of P is directly related to the number of bits used by B and Q.

Algorithm 5.3. A probabilistic version (PCPRD) of a CPRD is defined if Q is regarded as a classical device and if M performs probabilistic measurements. The QCPRD structure is modeled herein upon the visual cortical column (Verkhratskt, 2002, Kasai, 2003 and Miller, 2003), especially concerning the behavior of the device D and its relations with Q (Kasai 2003).

Remark 5.4: The evolution from **PCPRD** to **QCPRD** may account for the increase on cortical processing in nature.

5.3.2 Learning

The learning of the maps M_1, M_2 is implemented by beginning with large neighborhoods n_1, n_2 and pruning n_2 those connections at B and Q not used for P identification. If n_1, n_2 are initially very large, any P presented to R will be associated to a constant $f(R)$ and assumed as a known P. Now:

Proposal 5.2: The following are basic CPRD learning rules:

Rule 1: A rapid learning (imprinting) occurs if most (if not all) not used connections of M_1, M_2 are drastically ($n'_1, n'_2 \rightarrow 1$) pruned. In this condition, P is fixed and easily recognized in any other future occasion. Imprinting is efficient if the variability of P (v(P) in the closed interval [0,1]) is low, otherwise the number q of CPRD specialized to recognize $\{P_i\}_{i=1 \text{ to } r}$ increases as $r \rightarrow 1$ and v(P) $\rightarrow 1$.

Rule 2: A slow learning occurs if $n'_1, n'_2 \rightarrow \theta \ll r$ bits, the learning velocity being inversely proportional to θ. Slow learning is preferable if v(P) is large, since it allows the selection of the most significant s bits of P to be used in its recognition. However, the s $\rightarrow$ 0 as v(P) $\rightarrow$ 1. Again, different CPRD may be created to recognize sets similar $\{P_i\}_{i=1 \text{ to } r}$.

Spine pruning is demonstrated in both imprinting and slow learning in the zebra finch (Helmchen, 2002; Lieshoff and Bischo 2003). Sexual imprinting occurs mainly at the lateral neo-hyperstriatum (LNH) and slow learning at archineostriatum caudale (ANC). Both types of learning requires an initial increase of the number of spines in both (hormonally induced) LNH and ANC (environmentally induced). Spine density decreases in LNH within 2 days after exposition of to the female, whereas but requires around 3 weeks to stabilize in ANC. The density increase promoted by changing the animal from isolation to a social condition occurs in 3 days.

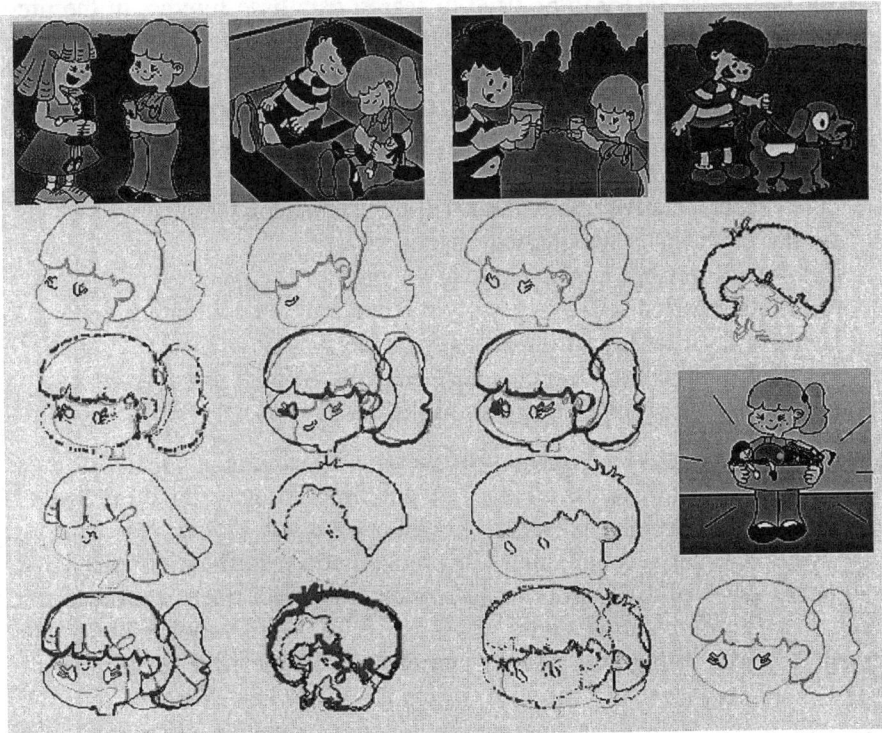

Fig. 5.6. Examples of PRCD recognition

5.3.3. Recognizing Faces

Algorithms **4** and **5** were implemented in the system **Sensor** that is being developed in our laboratory since 1997 (Rocha, 1997, www.eina.com.br/sensor) using the formalism of Fuzzy Formal Grammars to simulate natural vision systems. The efficacy of those algorithms in learning to identify defined images was tested using the Enscer Figure Data Base (www.enscer.com.br). This was done according to the following steps:

1. *select a searching element:* this is the element to be used to select the figures in the database. It is selected from one or more of the figures of this database. In the case of the example of Fig. 5.6, the chosen element was Laura, a female character;
2. *select the element features to be used in the search:* these are the components of the query to be used to search data base figures. In the present example, the query features were the face and hair contours;
3. *use the query elements for PCPRD training:* in the case of imprinting simulation it is sufficient to use just one example, whereas in the case of slow learning it is necessary to use examples from more than one training figure. These training examples are used to define the pattern of the query representative points, that will be used as the template to disclose similar elements in the queried images;
4. *test the efficacy of the system using another set of figures:* it is necessary to have a selected set of figures containing and not containing the query element, in order to test the efficacy of the system in correctly identifying the figures containing the query element, without missing any one or mistakenly taking other elements as the query one.

PCPRD attained an efficiency of 90% in cases similar to that of Juca and Laura's identification (Fig. 5.6) and imprinting simulations, when dendritic spine pruning preserved neighborhoods of 10% or less near the query representative points and the probabilistic identification was set to admit an error of 10%. Imprinting worked fine with query elements that are very prototypical as Laura. In the case of more variable query elements, slow learning was a better approach to characterize specific points of a small neighborhood variance, which are better key elements for a successful query. Such a characterization depends on successive pruning promoted by a given family of training examples. The PCPRD algorithm increased **Sensor** efficiency when compared to the previous contour identification algorithm in use, based on comparison of those points at which the contour experienced a significant change of direction (Rocha, 1997).

5.4 Fuzzy Logic and Conflict in $O(G|H,S_b)$

The ambiguity of $L(G \mid H)$ together with the distributed organization of $O(G \mid H, S_b)$ with respect to process, are certainly important sources for conflict among agents $n_i \in O(G \mid H, S_b)$.

Definition 5.5. Given

$d(s_o, s_t | H, S_b)$ and $d(\sim s_o, s_t' | H, S_b)$, $\mu(s_o, \sim s_o) = 0$, $s_t, s_t' \in V_t$ (5.7)

conflict occurs if

$$\mu(s_t, s_t') \rightarrow 1 \quad (5.8)$$

implying that if *confidence in* or *truth of* is defined as

$$\sigma(d(s_o, s_t | H, S_b)) = \mu(s_t, V_t), \sigma(d(\sim s_o, s_t' | H, S_b)) = \mu(s_t, V_t') \quad (5.9)$$

then

$$\sigma(d(s_o, s_t | H, S_b)) + \sigma(d(\sim s_o, s_t' | H, S_b)) > 1 \quad (5.10)$$

Proposal 5.3. DIPS conflict may arise from many causes, but mainly because distinct agents $n_i, n_j \in O(G | H, S_b)$ may shares pieces of information $d(s_i, s_j) \subset d(s_o, s_y)$ or $d(\sim s_o, s_t')$ collected at different moments t_i, t_j or for different sources n_r, n_s; they may use different tools or (V_n, P)s to process $d(s_o, s_t | H, s_i)$ and $d(\sim s_o, s_t' | H, s_j)$), etc. That is because two or more agents $n_i, n_j \in O(G | H, S_b)$ provide conflicting information when enrolled in solving a given task.

Theorem 5.1. The increase of the mean ambiguity $<\Omega(Ł(G | H, S_b))>$ of $Ł(G | H, S_b)$ enhances conflict in $O(G | H, S_b)$.

Proof: The mean ambiguity $\Omega(d(s_o, s_t) | H, S_b)$ of $d(s_o, s_t)$ is

$\Omega(d(s_o, s_t) | H, S_b) =$
$(\rho(d(s_o, s_t)) \log \rho(d(s_o, s_t)) - (1 - \rho(d(s_o, s_t))) \log(1 - d(s_o, s_t))$

$$<\rho(d(s_o, s_t))> = \frac{1}{m} \sum_{i=1}^{m} \rho(d_i(s_0, s_t | H, S_b)) \rightarrow 0{,}5$$

where m is the number of $d(s_o, s_t)$ accepted as belonging to $Ł(G | H, S_b)$ because $\rho(d(s_o, s_t)) \rightarrow 1$. Therefore

$$<\Omega(Ł(G | H, S_b))> \rightarrow m \text{ bits}$$

The increase of m augments the possibility that given two agents $n_i, n_j \in O(G | H, S_b)$ then

$$\rho(d(s_O, s_t | H, n_i)) \rightarrow 1, \rho(d(\sim s_O, s_t' | H, n_j)) \rightarrow 1,$$

and

$$\sigma(d(s_o, s_t \mid H, n_i)) + \sigma(d(\sim s_o, s_{t'} \mid H, n_j)) > 1$$

Conflict may be resolved by means of:

1. *Default Logic:* because matching priority supports derivation chains of the type

$$\rho(s_i, s_j) \to 0.5 \Rightarrow \alpha\, s_i\, \beta \to \alpha\, s_j\, \beta \text{ unless } \rho d(s_i, s_k) \to 1 \Rightarrow \alpha\, s_i\, \beta \quad (5.11)$$
$$\to \alpha\, s_k\, \beta$$

That is, given $\rho d(s_i, s_j) < \rho d(s_i, s_k)$, s_j rewrites s_i unless s_k is available to rewrite s_i. Here, the symbol $\Rightarrow$ means *to support*.

2. *Threshold Logic:* because it may assume that a sentence α s¡ β is allowed to be rewritten only if its number of available copies $q(s_i), q(s_j)$ are greater than a given minimum θ, that is

$$q(s_i), q(s_j) > \theta \quad (5.12)$$

3. *Temporal logic:* temporal reasoning may be supported by fuzzy languages, because $\rho(d(s_o, s_t))$ are subjected to temporal restrictions Δ, such that

$$\rho(d(s_o, s_t \mid H, n_k, \Delta) \to 1 \text{ only if } t < \Delta \quad (5.13)$$

and t is proportional to the number k = 2 of derivation steps of

$$d(s_o, s_1, \ldots, s_k, s_t).$$

As a matter of fact, the above logics may be considered as special cases in a broader family of logics, called

4. *Fuzzy Logic:* because it may assume that a sentence α s_i β is rewritten as

$$\rho(d(s_o)) = \overset{n}{\underset{i=1}{Q}} (\rho(d(s_o, s_t \mid H, n_i))) \quad (5.14)$$

where Q is a logical operator (e.g., *the majority of the rules, x of n rules, unless, etc.)*;

Remark 5.5. Many types of human reasoning are supported by the above logic (Rocha, 1992, Rocha, 1997). Many types of conflict are solved by means of consensus inspired in a logic of this type if

"most of the relevant pieces of information are true... then"

that is

$$\rho(d(s_o, s_t)) = \bigotimes_{i=1}^{n} (\rho(d(s_o, s_t | H, n_i) \otimes \sigma(d(s_o, s_t | H, n_i))) \to 1 \qquad (5.15)$$

However, not all types of conflict may be solve by the above techniques once the control of $\Pi(\rho(d(s_i, s_j) | H, S_b))$ is restricted by the power of the control language $Ł(C, | G, H, S_b)$ as stated by Theorems 2.5 and 2.6.

5.5 Recurrent Architecture Generates Entanglement Supporting Consciousness

A ubiquitous strategy used by large assemblies is signal reinforcement through recurrent processing. Computational loops correspond to circular sentences supported by $Ł(G| H, S_b)$.

Definition 5.7. A CD sentence $d([s_1... s_j ... s_c], s_t)$, $s_O = [s_1, ..., s_j,..., s_c]$ (Definition 4.1) is circular if there exists $d(s_1, s_j)$ such that $\rho\ (d(s_1, s_j)) \to 1$.
Forming loops is primarily a way to increase the computational weight of an assembly. Also, as a positive by-product, competition ends by increasing the coherence of the whole-brain activity, since the loops activate CDs and second-order catalysts. Therefore loops become critical factors for attention/consciousness. By increasing coherence they prepare the system for quantum computing, a resource that is able to overcome mismatching and destructive competition.

In the visual system, a sensory signal arriving at proximal dendrite sites of the pyramidal cortical neuron activates AMPA channels, depolarizes the membrane and displaces Mg^{2+} from the NMDA channel. The degree of this depolarization is assumed to be greater for tonic spiking state associated to wakefulness than for rhythmic burst firing observed during slow wave sleep. This AMPA activation encodes sensory information into a spike volley at the axon, which is distributed to other high order cortical neurons.

Later reentrant information from the last neurons is distributed over the distal sites of the pyramidal neuron dendrite and activates NMDA channels. If both the sensory information and the reentrant signal are coherently associated, then Ca^{2+} entrance is prompted and activates a stp, binding to CaM and then to CaMKII and other kinases, increasing the coherence of neuronal activity among all neurons that participate in the whole circuit

(Fig. 5.7). Such loops contribute to activate the coincidence-detectors and second-order catalysts, thus increasing the spatio-temporal coherence of brain activity, which is important for selective attention (LaBerge, 2001) and also to the putative generation of quantum entanglement among distributed agents, supporting complex cognitive functions and consciousness (Rocha et al., 2001).

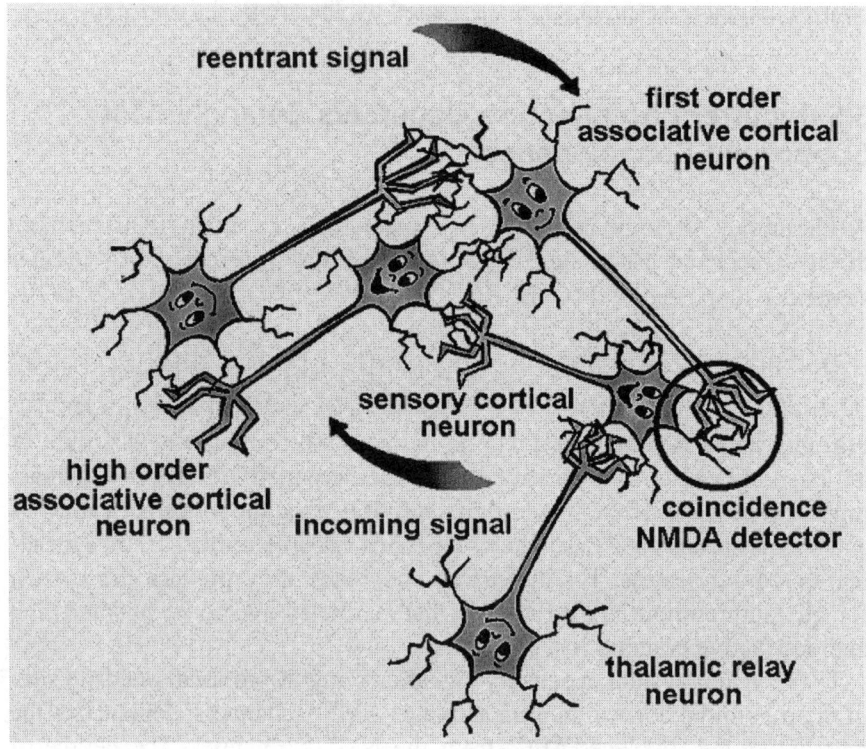

Fig. 5.7. **Reentrant loops in the brain:** *consciousness is entirely dependent on reentrant information*

One of the main characteristics of neuronal assemblies is the emergence of synchronous oscillations, in the theta and gamma frequencies (see a discussion in Pereira Jr. and Rocha, 2000), which are believed to increase the coherence among the cells belonging to the assembly, and to distinguish the assembly from its environment. By doing so, the assembly also prepares itself for quantum computing. The existence of convergence regions

5.5 Recurrent Architecture Generates Entanglement Supporting Consciousness

together with mechanisms of temporal synchronization of neuronal oscillations are important, but limited forms of integration. We have argued (Rocha et al., 2001) that possible and frequent conflict between specialized agents generates mismatches that could impair cognitive processing and context-related generation of adequate behavior. The putative and plausible existence of quantum computing in the brain would be a natural and scientifically-based alternative to explain how brain function could be integrated to such a higher level, and how a unified state of consciousness could be generated from a distributed parallel system.

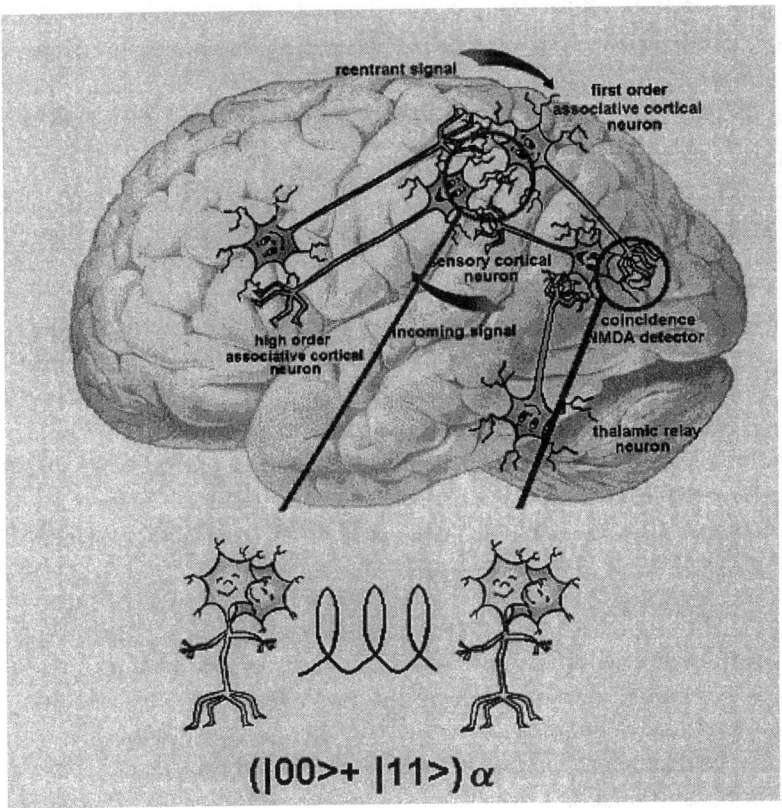

Fig. 5.8. Quantum processing in reentrant loops

Therefore, we propose that cognitive processing in the brain uses computational mechanisms that are very close to the strategies of parallel processing and non-local integration of information used in quantum computing. Entanglement in the whole brain is created and maintained by recurrent loops and temporal synchronization provided by the brain's reentrant architecture, e.g. over sensory and associative areas (Fig. 5.8).

The coherence-generating mechanisms involved in such processes – as the formation of loops, and temporal synchronization among several areas – are automatically impressed on Ca^{2+} dynamics. Considering that the action of electromagnetic forces upon a ion that is vibrating in the proper frequencies can change its internal state, then the action of such forces upon a population of ions can generate a inter-correlated change in the internal state of the whole population. This operation would mean that the ions can spontaneously *become entangled* for a brief period of time, in spite of the fact that the brain operates at physically high temperatures.

Another alternative is that quantum processing may be achieved by controlling the allosteric **(R,T)** states of proteins involved in coincidence detector stps ($d([s_1,...,s_c],s_t)$). In this condition, superposition, entanglement and quantum processing is assumed to be dependent on electronic states of molecules controlled by the entry of calcium through the NMDA channels and mR-Glu.

Of course, a population of Ca ions or Ca-interacting molecules that becomes entangled should decohere (reduce the superposed state by interaction with the environment) very fast. However, this tendency should not be used as an argument against quantum computing in the brain for three reasons.

First, because the brain possesses compartments where such populations can be functionally segregated for a period of time that would be sufficient to generate a conscious state.

Second, the most important reason is that *the waking brain is continuously generating new correlations* between Ca^{2+} ions and/or biological macromolecules, which can give continuity to conscious experience even when the correlations that generated the previous conscious state have vanished (of course, we know that *some* of these correlations are preserved through continued activity in *some* intra-cellular stps—corresponding to the phenomenon of memory).

Third, because decoherence is a necessary condition (for the consciousness-supporting quantum system) *to influence behavior,* as consciousness is assumed to do. By back-action on the biochemical systems where Ca^{2+} also participates (and has a central role), the interferences that occur during the entangled phase can have an influence upon which proteins are activated and how they are activated. Therefore, by considering this role of decoherence we can account for the apparent fact that consciousness is not an epiphenomenona but has a causal role, i.e., we can voluntarily determine a part of our behavior.

5.6 Superdense Codes

Amplitude interference of quantum states happens when amplitudes from different sources come together, since they may add in some places and subtract or cancel in others. Quantum computing takes advantage of interference to change state amplitudes for both writing and reading instructions (Gilmore, 1995). *Quantum superdense coding* (QSC) is the property that "in order to switch from any one of the four Bell states to all other four it is sufficient to manipulate only one of the two qubits while in the classical case one has to manipulate both" (Zeilinger, 1998).

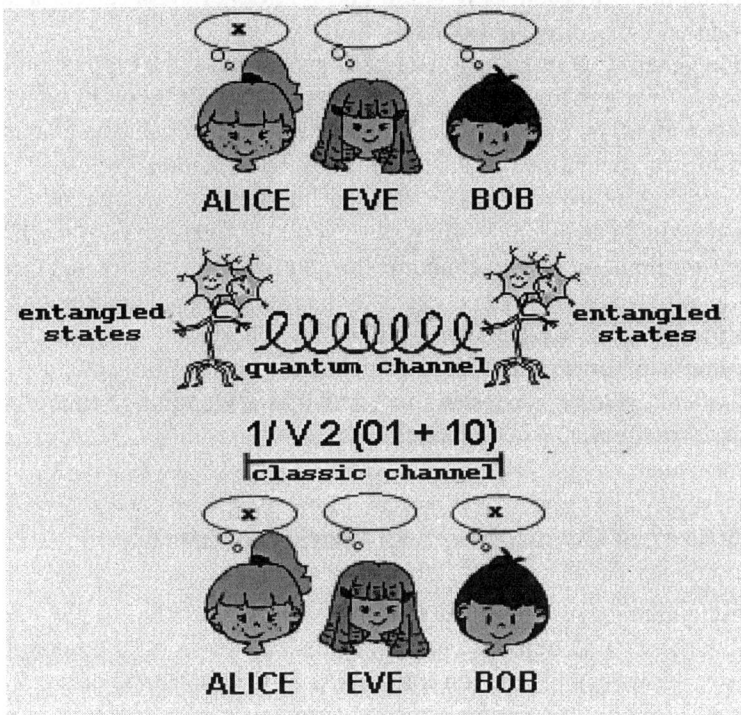

Fig, 5.9. Superdense codes: We do not mix information from different sensory channels.

A problem that QSC solves is to enable two protagonists, *Alice* and *Bob*, who share no secret information initially, to transmit a secret message **x** under the nose of an adversary *Eve*, who is free to eavesdrop on all their communications. This area of study has been called *quantum cryptography* (Fig. 5.9).

If Alice and Bob are limited to classical communication, they cannot detect eavesdropping. Now, if Alice and Bob's public communication is supplemented by a quantum channel, any eavesdropping disturbs quantum transmission in a way likely to be detected by Alice and Bob (Bennett and Shor, 1999). But channel noise also causes code corruption in quantum transmission. If the quantum channel is too noisy the best strategy (Bennett and DiVicenzo, 2000) is for Alice not to send the input qubit through the channel at all, but instead prepare a number of pure EPR pairs, and share them through the noisy channel with Bob, resulting in noisy EPR pairs. Then, using their ability to communicate classically, Alice and Bob distil a smaller number of good EPR pairs, and additional classical communication, to teleport the input qubit safely to Bob.

While generating an unitary state of consciousness, the brain should also separate the pre-processing of different 'qualia' up to the point when it all comes together. In the framework presented here, reliability of quantum processing in specialized brain areas would be obtained by using criptographic strategies to overcome undesired interference through noisy channels and provide coherent binding of quantum informational patterns. The strategy would be to keep a quantum communication channel together with classical axon-dendrite connections between such specialized regions.

Combining both kinds of long-range communication in the brain gives us a picture of brain function that is largely compatible with a popular model of information processing in contemporary cognitive neuroscience, the Working Memory (WM) model.

5.7 Quantum Computing and Working Memory

Neuroscientists have suggested that the prefrontal cortex might integrate spatial (where) and object (what) information, since cells responding to these two kinds of information intermixed in the prefrontal cortex engage in general-purpose temporary storage across many processing domains. Such lateral prefrontal function was named Working Memory, and assumed to be a conscious task. It is a short-term memory that is activated to support thinking processes, e.g. while performing a mental arithmetic task as the mental addition of 434 + 87. One common strategy is to add the units, tens and hundreds digits; while adding the tens it is necessary to keep other pieces of information in memory. This holding-in-memory while performing another task is the *modus operandi* of WM (Jonides, 1996).

Executive functions, in fact, seem to be spread across multiple regions of the frontal cortex, located on the outside surface of the lateral frontal cortex (LFC). However, the anterior cingulate cortex (ACC), located on the medial surface of the frontal cortex, is also activated in functional imaging studies of WM, and it is also considered an important executive area. This region receives important inputs from sensory systems and is connected to LFC. Both ACC and LFC are part of what is proposed as the frontal lobe attentional network, which involves also areas in the parietal cortex. This cognitive system is involved in selective attention, mental resource allocation, decision-making process, voluntary movement control, and/or resolving conflict between competing stimuli (LeDoux, 2002).

WM includes an "executive processor" and a "buffer". The "buffer" is supposed to be widely distributed over the brain, according to the contents of the information to be kept in WM. Finally, both LFC and ACC have important connections with the ventral frontal cortex (VFC), which in turn is one of the main entries of the neorcotical system in the emotional limbic system (LeDoux, 2002). In this way, emotions are automatically associated to WM contents.

There are limits of how much we can attend at one time. Some limitations are related to the similarity of the attended information. It is more difficult do attend to sources of information when both are presented to the same modality than when they are presented to different modalities. In a similar way a task of transforming a stimulus into a similar code, or dealing with similar semantic contents, is more difficult to attend to than when contents or codes are different (Posner, 1995). This may be explained by the required modality segregation by means of dense codes during quantum conflict solving.

Beyond this, there is a more general limitation on how much one can attend at one time. This general limitation can be demonstrated most clearly when all the specific sources of interference were removed. Perhaps because of these limitations, much of perceptual input goes unattended while some aspects become the focus of attention. Attending, in this sense, is jointly determined by environmental events and current goals and concerns. When appropriately balanced, these two kinds of inputs will lead to the selection of information relevant to the achievement of goals and lends coherence to behavior (Posner, 1995). This more general WM limitation may be explained by the limitations of maintaining long-lasting entangled circuits required to support the consciousness of WM contents.

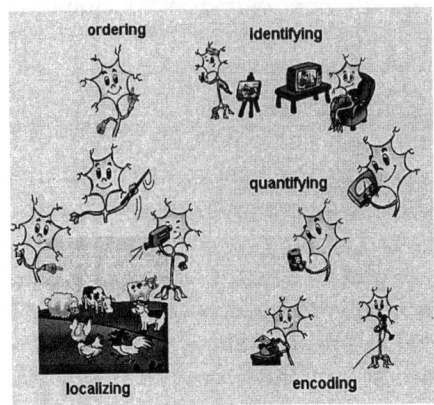 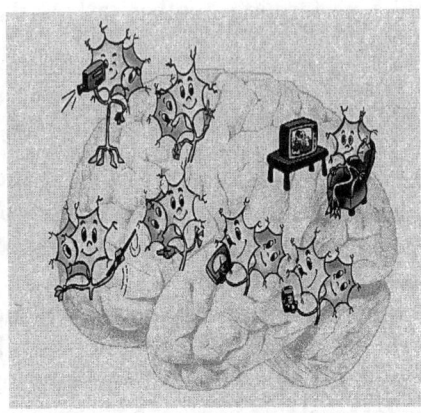

The task   *The entangled working memory*

Fig. 5.10. Working Memory: *a quantum computational space*

The existence of quantum communication channels between such areas, besides the classical connections, implies new properties to the WM function (Fig. 5.10). The processing of information in WM circuits cannot be considered to be purely serial, but includes parallel processing by several agents that are bound together through non-local integration. One of the consequences is that the role of the "executive processor" is relative, i.e., it is not the only agent able to apply the processing rules and integrate partial results.

6 Memetics and Cognitive Mathematics

The purpose of this chapter is to introduce the logic and basic assumptions of memetics, presenting its definition and presenting memetics as a potential answer to the problem of brain evolution and the human capacity of dealing with complex mathematical processes.

6.1 Memes

Humans are capable of imitation and so can copy from one another ideas, habits, skilled behavior, inventions, song and stories. These are all *memes*, a term which first appear in Richard Dawkins' book *The Selfish Gene* (Dawkins, 1976). In that book, Dawkins dealt with the problem of biological (or Darwinian) evolution as *differential survival of replicating entities* (Dawkins, 1976). By replicating entities Dawkins meant, obviously, genes. Then, in the final part of his book, Dawkins asked the question *'are there any other replicators on our planet?'* to which he answered *'yes.'* He was referring himself to cultural transmission and fancied another replicator – a unit of imitation (Blackmore, 1997; 1999). Dawkins first though of *'mimeme'*, which had a suitable Greek root (Dawkins' words) but he wanted a monosyllable word which would sound like 'gene' and hence the abbreviation of mimeme, or *meme*. A revolutionary new concept (actually, a truly Kuhnian paradigm shift) was born. Like genes, memes are replicators, competing to get into as many brains as possible.

One of the most important memes created by humans is the concept of *numbers*. Numbers are cultural inventions only comparable in importance to agriculture or to the wheel (Ifrah, 1985). Counting, however, is a process observable in a great number of non-human species. Millions of years before the first written numeral was engraved in bones or painted in cave walls by our Neolithic ancestors, animals belonging to several species were already registering numbers and entering them into simple mental computations (Dehaene, 1997). We, modern humans differ from the other species by being able to deal with numbers in a highly sophisticated manner, rather than the simple block-counting process characteristic of lower species, which is typically limited to 3 or 4 units.

The Oxford English Dictionary offers the following definition:

Meme: *An element of a culture that may be considered to be passed on by non-genetic means, esp. imitation.*

Memes can be thought as information patterns, held in an individual's memory and capable of being copied to another individual's memory. The new science of memetics is a theoretical and empirical science that studies the replication, spread and evolution of memes. As the individual who transmitted the meme continues to carry it, the process of meme transmission can be interpreted as a replication, which makes the meme a truly replicator in the same sense as a gene. Like the evolution of traits by natural selection of those genes that confer differential reproductivity, the cultural evolution also occurs by selection of memes with differential reproductivity, that is, those memes with the highest copying rates will take over those with lower copying rates.

Dawkins listed three characteristics for any successful replicators:

1. *copying-fidelity*: the more faithful the copy, the more will remain of the initial pattern after several round of copying;
2. *fecundity*: the faster the rate of copying, the more the replicator will spread; and
3. *longevity*: the longer any instance of the replicating patterns survives, the more copies can be made of it.

Just think of the example provided by Blackmore (1999), regarding the song 'Happy Birthday to You,' and you have a tremendously successful replicator, already copied (with high fidelity) thousand of millions of times (high fecundity) all over the world for several decades (longevity). In these characteristics, memes are clearly similar to genes and the new science of memetics imitates, to a certain extent, the example of genetics (a metamemetics phenomenon?).

Memetic and genetic evolution can interact in rich and complex ways, a phenomenon described as 'meme-gene-coevolution' (Blackmore, 1999). The study of cultural evolution is a branch of theoretical population genetics and applies population genetics models to investigate the evolution and dynamics of cultural traits equivalent to memes (Kendal and Laland, 2000). Research in gene-culture coevolution employs the same methods to explore the inter-relations of genes and cultural traits. Meme evolution may occur either exclusively at the cultural level, or through meme-gene interaction, both mechanisms having important consequences.

6.2 The Formal Meme

Let there be given two organisms:

$$O(G(H, S^i)), \Pi(\rho(d(s_i, s_j) \mid H, S^i_b)) \tag{6.1}$$

and $O(G(H, S^j)), \Pi(d(s_t, s_o \mid H, S^j_b))$

sharing the same grammar G, but having different knowledge K (see Definition 3.9)

$$K^i = \{d(s_o, s_t) \mid d(s_m, s_n) = d(s_o, s_t) \cap d(s_o, s_t) \neq \phi\} \tag{6.2}$$
$$K^j = \{d(s_o, s_t) \mid d(s_r, s_s) = d(s_o, s_t) \cap d(s_o, s_t) \neq \phi\}$$

about H because

$$\Pi(d(s_t, s_o \mid H, S^i_b)) \neq \Pi(d(s_t, s_o \mid H, S^j_b)) \tag{6.3}$$

This implies from Eq. 3.8 that there exists

$$d(s_m, s_n) = d(s_o, s_t) \cap d(s_o, s_t) \neq \phi, d(s_m, s_n) \in K^i \tag{6.4}$$

but $d(s_m, s_n) \notin K^j$

Definition 6.1. $O(G(H, S^j_b))$ reproduces (imitates) a sensory-motor process $d(s_o, s_t \mid S^i_b)$, as executed by $O(G(H, S^i))$, if it produces

$$d(s_o, s_t \mid S_b^i),$$

which reaches a final state similar to that reached by S^i; that is:

$$\mu(s_t \mid S^i_b, s_t \mid S^j_b) \to 1.$$

Theorem 6.1. $O(G(H, S^j))$ learns about $d(s_o, s_t \mid S^j)$, as executed by $O(G(H, S^i))$, if it is able to express $d(s_o, s_t \mid S^j)$ supported by the same action $d(s_o, s_t)$, reproduced by both $O(G(H, S^j))$ and $O(G(H, S^i))$. This learning occurs if $d(s_o, s_t)$ belongs to both $L(G(H, S^i))$ and $L(G(H, S^j))$.

Proof: As a consequence from the Definitions 3.8 and 6.1 and the fact that both $O(G(H, S^j))$ and $O(G(H, S^i))$ are able to execute $d(s_o, s_t)$ because it belongs to the languages expressed by both S^j and S^i.

Remark 6.1. Theorem 6.1 is about learning by imitation or by observation, a very important type of learning for some kinds of animal, including man. For instance, Petrosini et al. (2003) reviewed a series of papers disclosing the main properties of the means whereby the rat learns by observing its peers in solving maze problems. It is worth remarking that these authors also stressed the importance of the so-called "mirror neuron" (Rizzolati and Arbib, 1998) in supporting this type of learning. Here, the "mirror neuron" is assumed to be in charge of some of the derivations in $d(s_m, s_n)$, since G is distributed over the neurons of S_b.

Definition 6.2. $d(s_m, s_n)$ in Eq. 6.4 becomes a formal meme if $d(s_o, s_t)$ supporting it, is reproducible or imitable by another organism.

Theorem 6.2. The complexity (difficulty) in meme learning is directly proportional to

$$| \Pi(d(s_t, s_o | H, S^i_b)) - \Pi(d(s_t, s_o | H, S^j_b)) |$$

Proof: As a consequence of Definition 3.8f and Theorem 6.1.

Remark 6.2. Theorem 6.1 is also central to the concept of meme introduced by Dawkins (1976), since it set the basic conditions by which formal memes may spread (copied) in a population of $O(G(H, S))$s. In this way, meme dispersion is governed by the same evolutionary rules discussed in Chaps. 2 and 3. It is worthwhile to remember that those rules govern gene evolution, as well (Rocha and Massad, 2003b). Theorem 6.2 clearly defines one of the main restrictions on meme learning.

Proposition 6.1. $O(G(H, S^i))$ uses $d(s_o, s_t | S^i)$ associated to $d(s_o, s_t | S^i)$ in order to induce $O(G(H, S^j))$ in reproducing $d(s_o, s_t | S^i)$ and to recover $d(s_o, s_t | S^i)$ as $d(s_o, s_t | S^j)$. In this way, $O(G(H, S^i))$ signals $d(s_o, s_t | S^j)$ to $O(G(H, S^j))$ using $d(s_o, s_t | S^i)$.

Proof: Follows from Definitions 3.8f, 6.1, 6.2, and Theorem 6.1.

Remark 6.3. Meme learning has generally been characterized as a decision by the organism S^j, resulting from observation of the behavior of a knowledgeable organism S^i. Proposition 6.2 shows that meme spread may also be due to a decision by S^i to promote the learning of a meme $d(s_m, s_n)$. In this way, the meme spreading may also be supported by instruction.

6.3 Improving Meme Spread

Let S and M be such that is possible to guarantee initially that there exists

$$S_i, S_j \subset S \text{ and } M_i, M_j \subset M \tag{6.5}$$

and

$$d(s_o, s_t \mid S_i, M_i), \rho(d(s_o, s_t \mid S_i, M_i)) \to 1, \qquad (6.6)$$

$$d(s_o, s_t \mid S_j, M_j), \rho(d(s_o, s_t \mid S_j, M_j)) \to 1$$

such that

$$d(s_m, s_n) = d(s_o, s_t \mid S_i, M_i) \cap d(s_o, s_t \mid S_i, M_i), \qquad (6.7)$$

$$\rho(d(s_i, s_j)) \to 0.5.$$

In this condition

Theorem 6.3. $O(G©(H, S^i))$ is able to learn to signal about $d(s_o, s_t)$ supported by $d(s_o, s_t \mid S_i, M_i)$ using $d(s_o, s_t \mid S_j, M_j)$.

Proof: From conditions 6.5 to 6.7 there always exists

$$d(s_m, s_n) = d(s_o, s_t \mid S_i, M_i) \cap d(s_o, s_t \mid S_i, M_i),$$

$$\rho(d(s_i, s_j)) \to 0.5.$$

Given that G is a self-controlled grammar, it is possible to evolve $s_c \in C$ (see Definition 2.5 and Eq. 2.34) such that $\rho(d(s_i, s_j)) = f(q(s_c))$. In this condition it is possible to $\rho(d(s_i, s_j)) \to 1$, such that $d(s_o, s_t \mid S_j, M_j)$ becomes expressible as $\rho(d(s_o, s_t \mid S_j, M_j)) \to 1$ as a consequence from $\rho(d(s_i, s_j)) \to 1$.

In this way, $O(G(H, S^i))$ may use $d(s_o, s_t \mid S_j, M_j)$ to sign $d(s_o, s_t)$ supported by $d(s_o, s_t \mid S_i, M_i)$.

Remark 6.4. The language $Ł(G(H, S_j, M_j))$, which is composed of all $d(s_o, s_t \mid S_j, M_j)$ used to signal about $d(s_o, s_t)$ composing $Ł(G(H))$, is the type of human-like language that may evolve in G. If S_j, M_j signify the phonetic motor system, then $Ł(G©(H, S_j, M_j))$ is an oral language; otherwise it is a signed language.

Theorem 6.4. The expressiveness $\theta(Ł(G(H, S_j, M_j)))$ increases as the cardinality of S_j, M_j is augmented by evolution.

Proof: as a consequence from Eqs. 2.22 to 2.5 and 6.5 to 7, Theorems 2.1, 2.2, 2.5, 6.1, and 6.3.

Remark 6.5. As recently proposed by Hauser et al. (2002), the evolution of human language is a consequence of the evolution of the brain for more general purposes. Also, according to recent findings published by Enard et al. (2002), the complexity of human language is mainly due to the increase of the complexity of oral-pharyngial motor control. The increase of the $\theta(Ł(G(H, S_j, M_j)))$ in the case of human languages allowed another mechanism for meme spreading, based on the rules defined by Proposition 6.1, that we used to call *teaching*.

6.4 Memetic Channels and the Brain

Communication among **DIPS**' agents is established by means of two main strategies:

1. *Mail addressing*: both the sending and the receiving agents know themselves, that is to say they have the capacity to address messages specifically to each other. Imitation is mainly a mail-addressing memetic channel, since it is based in individual "contacts."
2. *Broadcasting*: agents deliver messages that are not specifically addressed to another defined agent, but to those interested in the subject. Instruction is the first mechanism to implement a broadcasting memetic channel since the action of one organism is communicated to a group of peers. Language, by supporting teaching, increases the capacity of a broadcasting memetic channel. Writing widely broadens this capacity.

Since, unlike genes, memes do not come packaged with instructions for their replication, our brains provide this function, strategically guided by a fitness landscape that reflects both internal drives and a world-view that is continually updated through meme assimilation (Gabora, 1997).

The evolution of the human brain began with the macro evolutionary events that culminated with the first species of the Homo genus, about two and a half million years ago. By about 160,000 years ago *H.sapiens* (Conroy, 1997; White et al., 2003) had brains as large as ours and the other sapiens sub-species, the Neanderthals had brains with a volumetric capacity larger than ours. They controlled fire, had cultures and probably had some form of language as well. The evolution of the brain is achieved by both increasing the number of its neurons (brain size increase) and by allowing new specializations for these new neurons.

The increase in brain size, however, had a price (Blackmore, 1999). Oversized brains are expensive to run and ours consumes 20% of the body's energy for a mass corresponding to only 2% in weight. In addition, brains are expensive to build. The amount of protein and fat necessary for the development of the human brain forced the first members of the Homo genus to increase their meat consumption, which entailed better hunting strategies, which in turn fed back to increased brain size. Finally, big brains are dangerous to produce. The increase in brain size, along with the bipedal gait of the Homo species resulted in severe birth risks. Big brained human babies have enormous difficulty passing through the birth canal. In addition to higher maternal and fetal mortality, this results in the human baby being born prematurely, as compared with other primates. On one hand, this has the beneficial consequence that our brains have greater neuronal plasticity, which increases their learning capacity. On the other hand, the complicated twisting maneuvers the human fetus has to do in order to pass through the birth canal means that the human female rarely is able to deliver without assistance. This also contributes to socialization and additional selection for brain growth.

Our brains have changed in many ways other than just size. The modern human prefrontal cortex, oversized when compared with other hominids, is fed by neurons coming from practically all other parts of the brain. Its role in the complex cognitive abilities of modern humans is still to be fully understood, but we already know that when damaged by accident or surgical removal (a common practice some decades ago) the victim is severely limited in calculation performance (in addition to personality changes). However, this size increase may free the brain to *invent* new types of neurons, by modifying gene expression, whenever an increasing social selective pressure appears.

6.5 The Evolution of our Mathematical Competence

Let us imagine the African environment of about 150,000 years ago. It is now widely accepted that our species evolved in a lake environment at about this time (Fig. 6.1). The first humans began to organize themselves into small groups of hunter-gatherers, with few contacts outside their own clans. With the abundant supply of animal protein and fish fat, their brains achieved the size and organization observed in current members of our species. However, their numerosity was probably restricted in the same ways as that seen in modern hunter-gatherer societies. As those primitive human groups started out on their journey toward the northern parts of Africa (and, mainly after the out-of-Africa migration waves towards the Caucasus and then Europe) they began to come in contact with other groups of primitive humans, the Neanderthalers, who had been dwelling in Europe for some 100,000 years at that time. The clash of cultures had well known disastrous effects for the latter with our own sub-species prevailing and the disappearance of our cousins some 30 to 50 thousand years ago.

124   6 Memetics and Cognitive Mathematics

Fig. 6.1. The primate hominides: *it started around the lakes ..*

It was not, however, until the first nomadic tribes of humans settled in communities, following the agricultural revolution some 10 thousand years ago, that trade between different groups and/or individuals pressed for the development of a more sophisticated numerical system. Trade, and the necessity for book-keeping, was also a selective pressure for the evolution of mathematics beyond simple counting. In our own country, it is possible to find evidence of such a process.

The archeological site known as *Pedra Furada,* in the northern part of Brazil, is one of the richest sites of primitive human groups (Guidon, 1998), aged between six and twelve thousand years (as in any other dating, those figures are subject to ferocious argument within the archeological community). Its archeological findings are currently divided into two cultural traditions, named the "Nordeste" tradition and the "Agreste" tradition.

The Nordeste tradition had two phases, the *Pedra Furada* and the *Pedra Talhada.* It is characterized by pictorial (humans, animals and vegetables) and non-pictorial graphics (Fig. 6.2). The pictorial graphics represent action, as a general rule, and are more prevalent than the non-pictorial graphics. Human figures are represented with several cultural attributes and are depicted in their daily life activities. Four main themes can be identified: dancing, sexual activities, hunting, and activities around a tree. One specific figure, a non-realistic picture dated from around nine thousand years ago, appears repeatedly in other archeological sites of the northeastern part of Brazil. The styles of the Nordeste tradition are named *Serra da Capivara* (SC), which is the oldest, *Serra Talhada* (ST) and, most recently, *Serra Branca* (SB). The ST, which is intermediate in time, is the most complex, denoting an evolutionary process from SC to ST, but a regression in style to SB. This process is believed to have happened around six thousand years ago.

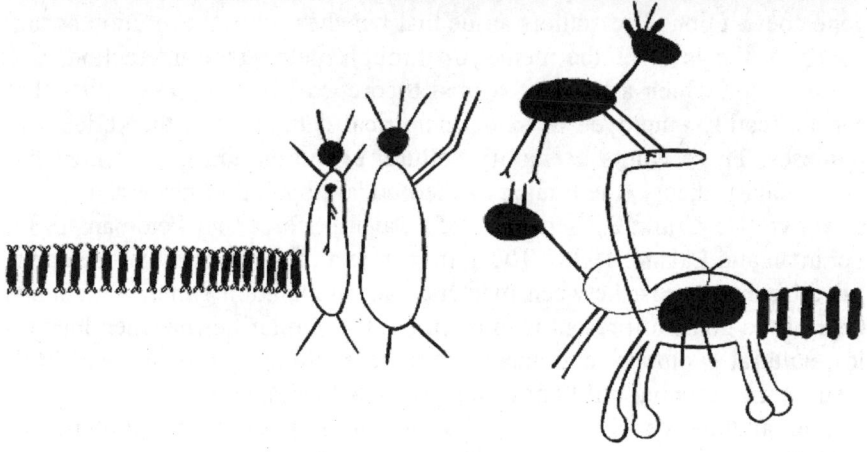

Fig. 6.2. The paintings of Pedra Furada: *are they depictions of early arithmetic?*

The Nordeste tradition disappeared, probably due to a combination of climatic changes and the arrival of better warriors groups, which forced the former inhabitants to abandon the area.

The Agreste tradition is characterized by human graphics and rare animals. Graphics representing action are rare and are restricted to hunting. Pure graphics are more prevalent than in the Nordeste tradition. Frequently, the Agreste graphics were done inside the Nordeste panels, but can be distinguished from these by their lower technical quality. The Agreste and Nordeste pictures coexist in the area by about 10 thousand years but the Agrest can be disappeared some 4-3 thousand years ago.

6.6 How Memes Spread

Several attempts have been made to provide the new science of memetics with a mathematical framework for modeling the spread of memes. *The Journal of Memetics* (http://jom-emit.cfpm.org/all.html), an electronic journal dedicated to memetics, presents a number of articles dealing with the mathematics of memetics. The great majority (if not the totality) of these words are essentially adaptations of population genetics. In the paper by Edmonds (1998), for instance, a classification of memetics models is presented, with an interesting discussion of the possibilities of modeling memetics. Kendal and Laland (2000) discuss the phenomenon of meme-gene coevolution. The authors argue that whether cultural evolution occurs purely at the level of the meme, or through meme-gene interaction, is a question for which a body of formal theoretical work already exists that can be readily employed to model empirical data and test theoretical hypotheses. These works exemplify cultural evolution and gene-culture coevolutionary theory, the branch of theoretical population genetics referred to above (Boyd and Richerson, 1985; Cavalli-Sforza and Feldman, 1981; Feldman and Laland, 1996). The authors reject the argument that meaningful differences exist between memetics and population genetics methods. One of the goals of present is to point out the similarities between memetics, cultural evolution, and gene-culture co-evolutionary theory, and to illustrate the potential utility of genetic models to memetics.

The authors conclude that cultural evolution and gene-culture co-evolutionary modeling paradigms can be effectively employed to enhance the quantitative study of memetics. Simple and complex cultural phenomena such as behavior patterns, belief systems, and institutions can be analyzed by characteristics of associations between easily definable and quantifiable memes. The quantitative approach can be used to describe meme diffusion dynamics, and make sense of patterns of variation in memes. The methods can also clarify why and how human attributes evolved in conjunction with memes, how they continue to evolve, and what is the basis of any stability or maintenance of the trait

In another interesting model, Gatherer (2001) shows simple computer simulations of the interaction of genetic factors and memetic taboos in human homosexuality. These simulations clearly show that taboos can be important factors in the incidence of homosexuality under conditions of evolutionary equilibrium, as for example in states produced by heterozygote advantage. However, frequency-dependent taboos, i.e. taboos that are inversely proportional to the incidence of homosexuality, cannot produce the oscillating effect on gene frequencies predicted by Lynch (1998). Effective oscillation is only produced by rapid withdrawal and re-imposition of taboos in a non-frequency-dependent manner, and only under conditions where the equilibrium incidence of homosexuality is maintained by heterozygote advantage, or other positive selective mechanism. Withdrawal and re-imposition of taboo under conditions where homosexuality is subject to negative selection pressure, produce only feeble pulses, and actually assist in the extinction of the trait from the population. Additionally, it is shown that frequency-dependent taboos assist in a more rapid achievement of equilibrium levels, without oscillation, under conditions of heterozygote advantage. An attempt is made to relate the simulations to past and contemporary social conditions, concluding that it is impossible to decide which model best applies without accurate determination of realistic values for the parameters in the models. Some suggestions for empirical work of this sort are made.

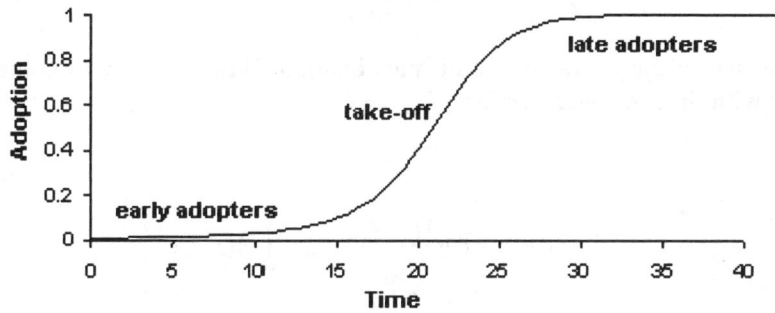

Fig. 6.3. Logistic model for innovation spread

The above discussion is hence related to the population genetics approach to memetics. Let us see now an ecological approach—that is, how to model meme spread by dynamical systems. As a matter of fact, the first approach to the spread of ideas by dynamical systems was that due to Rogers (1995), whose classical book, the *Diffusion of Innovations*, is a landmark of modeling the spread of ideas and concepts. Actually, when Rogers wrote the first edition of his book forty years ago (Rogers, 1962), the very concept of meme has not been proposed yet. In its fourth edition (Rogers, 1995), however, this book still misses the idea of a meme. Notwithstanding this fact, the spread of innovations is not an alternative to memetics dynamics. The so-called logistic model of innovation spread (Fig. 6.3) is one way of simulating this dynamics.

In this model, it is assumed that, from a total population n, a fraction a adopted a given novelty. Therefore, there remains a fraction $n - a$ individuals ''susceptible` to the innovation. The model also assumes a contact rate λ between 'infected' and 'susceptible' individuals.

The rate of growing of adopters is therefore given by:

$$\frac{da(t)}{dt} = \lambda a(t)(n - a(t))$$

which can easily be solved as:

$$a(t) = \left[1 + \frac{(1-a_0)}{a_0} \exp(\lambda t)\right]^{-1}$$

This is one of the forms of the logistic equation. When the rate of contact varies with time, the equation is:

$$a(t) = \left[1 + \frac{(1-a_0)}{a_0} \exp\left(\int_0^t \lambda(s)ds\right)\right]^{-1}$$

The reader familiar with the mathematical theory of epidemic spread will easily recognize the above model as an epidemiological model in a new disguise. It is indeed a model of spread which could be made more sophisticated in order to take into account other facts related to transmission, like the reproduction rate of ideas spreading.

An interesting example of the spread of a new meme is represented by the case of hybrid corn in Iowa farms in the late 20's and early 30's, described by Rogers (1995). At the time, farmers chose the best seeds from a given year's yield for the following year's planting. The hybrid corn seeds, in contrast, besides giving a greater yield, were handicapped by the need to buy new seeds every year. The example of hybrid corn is described in detail by Rogers. In the following analysis, we revisit the spread of hybrid corn in Iowa, applying an original model.

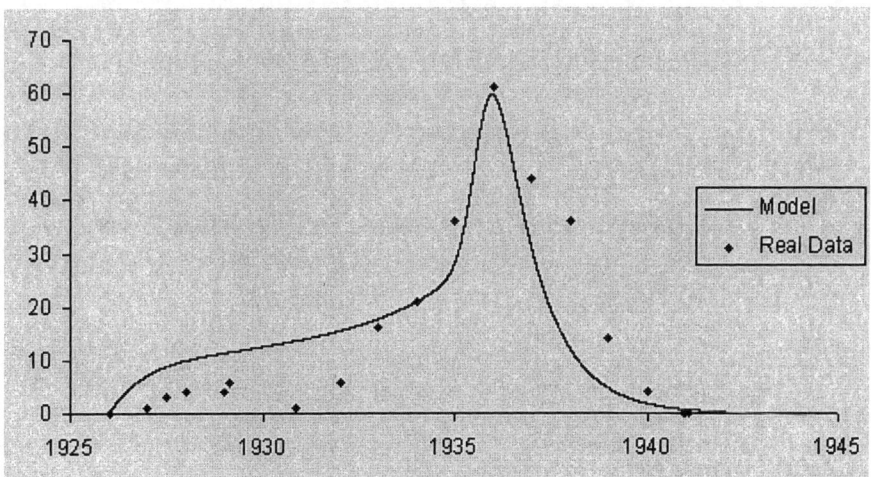

Fig. 6.4. Modeling hybrid corn technology diffusion: *Number of farms per year adopting hybrids*

The model assumes two types of farmers, called "susceptible" to the innovation, denoted S_1 and S_2, one of them more opinionated than the other. Both susceptible types of farmers are subject to broadcast advertising, and adopt the innovation at a rate of λ farmers per time unit in this way. Once having adopted the innovation, susceptible farmers pass to a new state, called "infected" by the innovation, denoted I_1 and I_2, depending on the previous states, whether S_1 or S_2, respectively. In addition to the broadcast advertising, farmers could adopt the new meme by a kind of "infectious" contact with the farmers who had already adopted the innovation. This occurred at rates β_1 and β_2 potentially infectious contacts per time unit, depending on the contact states of S_1 or S_2, respectively. Once the meme was adopted, the farmers were removed from the infectious state to a new, "resistant" state, denoted **R₁** and **R₂**, from the states I₁ and I₂, respectively, with rates γ₁ and γ₂. The more opinionated farmers, I₂, influence the more susceptible farmers S₁, through a new contact rate r, but are, in turn, susceptible to innovation by contact with farmers from the class I₁. The model assumes also that the broadcast advertising rate λ decreases with time according to a logistic function, $\lambda(t) = 1/(1+\exp[-\kappa t])$ and that the direct contact rates β_i, i = 1, 2, increased according to another logistic function, $\beta_i(t) = 1/(1+\exp[\omega_i t])$. The model's dynamics are described by the following set of ordinary differential equations:

$$\frac{dS_1}{dt} = -\lambda(t)S_1(t) - \beta_1(t)S_1(t)I_1(t) - rS_1(t)I_2(t)\theta(I_2(t) - 0.2)$$

$$\frac{dS_2}{dt} = -\lambda(t)S_2(t) - \beta_2(t)S_2(t)I_1(t)$$

$$\frac{dI_1}{dt} = \lambda(t)S_1(t) + \beta_1(t)S_1(t)I_1(t) + rS_1(t)I_2(t)\theta(I_2(t) - 0.2) - \gamma_1 I_1(t)$$

$$\frac{dI_2}{dt} = \lambda(t)S_2(t) + \beta_2(t)S_2(t)I_1(t) - \gamma_2 I_2(t)$$

$$\frac{dR_1}{dt} = \gamma_1 I_1(t)$$

$$\frac{dR_2}{dt} = \gamma_2 I_2(t)$$

The results of the numerical simulation of the model can be seen in Figure 6.4, which demonstrate the good fitting capacity of this model with respect to real data from the Iowa farmers.

We also calculated the basic reproduction ratio of the innovation, the threshold number of "infected" farms, below which the innovation would not spread to the other farms. This is a parallel to the basic reproduction ratio, R_0, (Anderson and May, 1991) of infections and it is considered the key parameter related to infectious dynamics. For the proposed model, the threshold condition is given by λ = 0; that is, provided the broadcast advertising is positive, the innovation will spread. When this is not the case, that is, when the spread is dependent only on the contact between farmers who adopted the new meme and those who are still susceptible, the threshold condition is given by

$$R_0 = \frac{\beta_1}{\gamma_1} + \beta_2 \frac{r}{\gamma_1 \gamma_2}$$

As the rates λ, β_1, and β_2 are time-dependent, this parameter is variable with time, starting with zero (when λ is different from zero) and growing to more than 8000 at the peak of the "epidemic" when calculated with the values used in the simulation of the model.

6.7 How the Number Meme Spread

According to Rogers (1995), the characteristics of innovations, as perceived by individuals, which help to explain their different rates of adoption are:

1. *Relative advantage*, the degree to which an innovation is perceived as better than the idea it supersedes;
2. *Compatibility*, the degree to which an innovation is perceived as being consistent with the existing values, past experiences, and needs of potential adopters;
3. *Complexity*, the degree to which an innovation is perceived as difficult to understand and use;
4. *Triability,* the degree to which an innovation may be experimented with on a limited basis; and
5. *Observability*, the degree to which the results of an innovation are visible to others.

Therefore, innovations that are perceived by individuals as having greater relative advantage, compatibility, triability, observability, and less complexity will be adopted more rapidly than other innovations.

Now, returning to the main theme of this chapter—the spread of the number meme—we may imagine a scenario from about 10,000 years ago, when trading was starting to blossom and the 'natural' capacity of numeracy, that is, counting in blocks up to four or five, was not enough anymore. Those individuals with intelligence slightly above the average of the time could invent counting procedures and techniques superior to KFN. Those new techniques that were perceived by others as having greater relative advantage, compatibility, triability, observability, and less complexity certainly spread rapidly, representing, therefore, a paradigm shift in the current culture of the time.

It is easy to conclude that a growing arithmetic capacity in a local group of individuals represented a greater relative advantage at a competitive stage of the primitive trading system. The new way of dealing with numbers should be easily tried, the results readily observable and, if the system evolved in a gradual way, the complexity should be built up on each new step, in order not to be a restrictive factor.

It may, therefore, be concluded that the evolution of mathematics, as known today, started with the spread of a new meme (or rather a set of new memes) in a manner rather like the way described by Rogers (1995), that is, for each step, early adopters learned by imitation, followed by a take-off phase of quick spread, reaching eventually an equilibrium with the late adopters. The theory of memetics, along with the theory of innovation diffusion, explains in a rather plausible way such evolutionary events which changed human culture forever.

Finally, it is worth mentioning the curious fact that the evolution of number systems very similar to each other, occurred in an independent manner in places far away in time and space, such as Egypt and Guatemala! This indeed demonstrates the power of new mathematical ideas in providing their inventors and adopters with a great competitive advantage.

6.8 Future Perspectives in the Memetics of Cognitive Mathematics

So far, we have addressed the hypothesis of the evolution of our mathematical capacity based on plausible assumptions of meme-gene coevolution of brain size and complexity. It is possible, however, that the evolution of mathematical ability is more dependent on a more active form of meme transmission, namely, teaching. If this is the case then the development of mathematics beyond the **KFN** level occurred later in our evolutionary history. Teaching is a more sophisticated form of meme transmission than simple imitation. In addition, teaching is a kind of social interaction, which occurs in more complex communities. It may have evolved as a particular case of altruism. Teaching is a form of giving someone else precious information that could result in differential reproductivity in the receiving individual. Therefore, teaching probably evolved by the same mechanisms currently accepted for the evolution of altruism: kin selection or reciprocation. In the former, the individual who teaches some useful trick to another individual increases his/her inclusive fitness if the receiver shares a substantial number of genes with him/her. In the latter mechanism, the altruistic teacher gives his/her precious information with the expectation that in later interactions the current receiver will share other kinds of information with him/her. Both mechanisms make sense from the gene and meme points of view. A third mechanism could also be involved in the transmission of a mathematical memeplex: selfish teaching. In this form of 'altruism,' the teacher shares information with the receiver based on the belief that this will increase his/her own survival or the survival of the group. The meme-gene coevolution aspect of this kind of meme transmission, however, should occur under the setting of group selection. Group selection, although still controversial among population geneticists, may indeed occur in cultural evolution, as also discussed by Blackmore (1999).

The important point we would like to emphasize is that mathematical memeplexes can be transmitted, either by imitation or by teaching, and that both mechanisms can be inferred in the evolution of our cognitive mathematical abilities, and through the same meme-gene coevolutionary mechanisms.

Another area that could be subject to future investigation related to the meme-gene coevolution of mathematical memeplexes would be the development of mathematical tools to deal with the dynamics of meme transmission and the evolutionary aspects, much in the flavor of the neo-Darwinian synthesis. Population genetics is a mathematically rich area of evolutionary biology and perhaps one could propose a new specialty that could be called population memetics in which the adaptive value of meme transmission could be quantified and predictions on meme dynamics could be made. But this also is matter for future works.

Mathematics is a tremendously rich collection of ideas, concepts, tools, etc., that can be characterized as a memeplex. It is always evolving and, in addition to its obvious role in changing the world, and our world vision, has certainly helped in the evolution of our cognitive capacity. Since the first hominids surpassed the imitation threshold, individual mathematical memes more complex than our innate subtizing (in addition to all other memeplexes that characterize human culture), have been transmitted by imitation and later on, by teaching, creating an autocatalytic virtuous circle that culminated in the human brain.

7 Modeling of Arithmetic Reasoning

In this chapter we extend the scope of an earlier model (Rocha and Massad, 2002 and 2003a), wherein innate cerebral circuits were proposed to deal with numbers, and where some of these circuits were specialized in dealing with distinct types of number. On this view, the arithmetical process is distributed among several brain areas, and different strategies are used to solve the same arithmetical calculations. It is also assumed that some important properties of the counting system may be disclosed by studying the relation between calculation time and the size of numbers.

7.1 Counting

Counting is a process dependent on a distributed representation of number in the brain (e.g., Butterworth, 1999; Dehaene, 1997; Dehaene, 2002). The efficacy of the processing is mainly dependent on the optimization of the control of the eyes and hands in attending the objects to be counted (e.g., Butterworth, 1999; Massad and Rocha, 2002, Rocha and Massad, 2003a). Counting is also proposed to be the underlying process for arithmetic calculation since it has been demonstrated that the (response) time required to produce the result of an arithmetic calculation is dependent on the size of the relevant numbers (e.g., Ashcraft, 1992; Dehaene, 1997; Fayol, 1996; Hinrichs et al. 1991; McCloskey, 1991; McCloskey et al., 1991; Siegler, 1996). It has also been demonstrated that different strategies may be used by the same person to solve the very same calculation (e.g., Siegler, 1996). Despite these advances, no formal model other than the *triple-code model* proposed by Dehaene and colleagues (Dehaene et al. 1998, Cochon et al. 1999) has been proposed so far to describe the actual experimental data on brain arithmetic. The triple-code model assumes that numbers are represented in the human brain in three distinct formats: as Arabic numerals, as sequences of words, and as analog representations of the corresponding numerical quantities. Also, arithmetic calculation is mostly dependent on verbal encoding.

Because of the foregoing considerations, a new class of fuzzy numbers, called **K** Fuzzy Numbers (**KFN**), was proposed by Rocha and Massad (2002) to implement arithmetical knowledge in a Distributed Intelligent Processing System (**DIPS**). The proposed system is designed to simulate brain function and experimental data about arithmetic capability in both man and animals. KFN's main property is the dependence of the size of the base of its membership function $\mu_{di}(\sigma)$ (Pedrycz and Gomide, 1998) on the value of the number d_i encoding σ.

Another interesting KFN property allows for many different solutions to be tried on the very same arithmetic problem. Indeed, this seems to be the strategy employed by the brain in the case of arithmetic processing. Many studies of the means whereby children in different countries solve standard arithmetic problems have revealed that youngsters use multiple strategies (Butterworth, 1999; Dehaene, 1991, Dehaene, 1997; McCloskey et al., 1991; Siegler, 1996):

1. *total manipulation*: the child counts separately each set to be processed by pointing, marking, etc. each of their elements, and then counts by the same process each element of the union, or the complement, etc. of these sets to get the final result;
2. *simplified manipulation*: the result is obtained by counting each element of the union, or the complement, etc. of the sets to be processed;
3. *optimized manipulation*: the result seems to be obtained by performing the minimum counting, which varies according to the type of calculation to be performed, and
4. *mental calculation*: the result is quickly processed by specialized circuits.

Also, the type of function explaining the effect of number size, as reported in the literature, correlates with the kind of manipulation used (e.g., McCloskey et al., 1991). Finally, training changes the frequency of use of the different types of strategy, such that adults tend to rely more on mental calculation than do children (Siegler, 1996).

7.2 A Model for Number Sense

Counting, as for example a collection of animals on a farm (Fig. 7.1), is a very complex process requiring the participation of sets of neurons located in different areas of the brain, with each set in charge of solving one particular segment of the entire process. These sets are specialized to collect sensory data in order to recognize the object to be counted, to perform calculations, to keep track of the counted objects, and to speak about the results. To count is, therefore, a task for a **DIPS**.

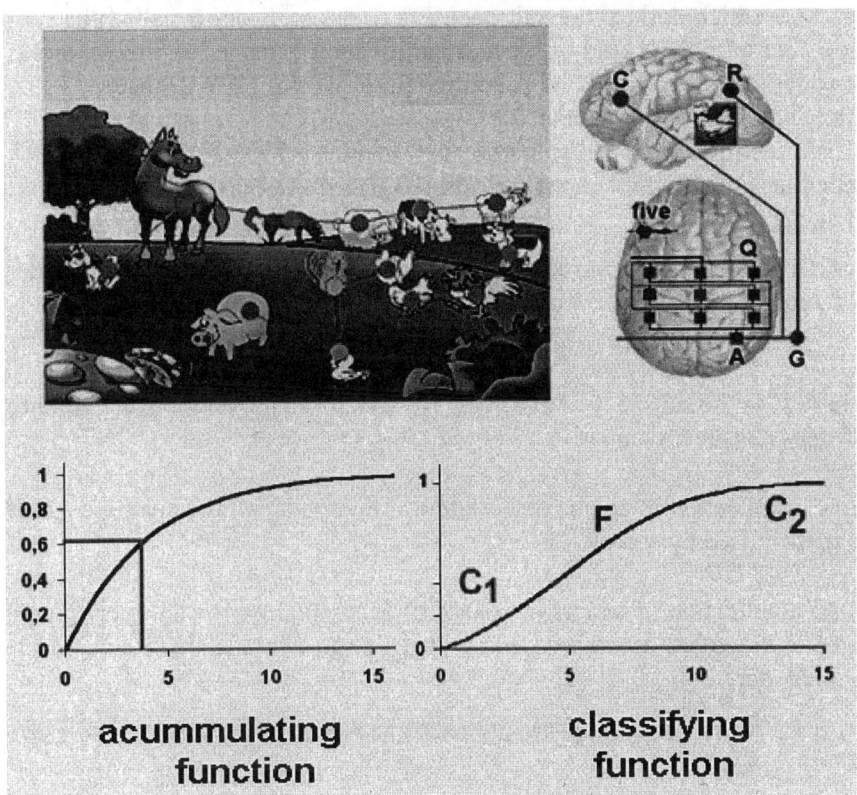

Fig. 7.1. The KFN counting circuit: *an ancestral neural circuit for counting.*

7.2.1 Identifying numerosities

Let the counting DIPS labeled KFN (Fig. 7.1) consist of two sets of agents:

S: collection of sensory agents s which measure (m) variables **v** in U as $m = f(v)$, in a given subspace $f \subset U$; and

R: a collection of recognition agents specialized for identifying objects $o_i \in U$, by means of a set Ψ_i of relations between the above measures m_i of these o_i. The image $I(o_i)$ used by R to identify o_i is the minimum set of the relations uniquely associated to o_i and no other o_j in U; or $I(o_i) \subset \Psi_i$.

Many objects o_i may be simultaneously identified by R because of the redundancy inherent to any DIPS. The maximum number β_i of I (o_i) simultaneously identified by R is dependent on and limited by the agent redundancy r, which defines the subspace $F = \{f_i\}_{i=1 \text{ to } r}$, and is called the *sensory field* of S. This kind of block quantification is named *subtizing*, and is considered a special kind of counting (Butterworth, 1999, Dehaene, 1997, Fink et al. 2001; Piazza, et al. 2002).

The identification of the cardinality of any set O_i of objects o_j in U, called here *cardinality quantification* (**CQ**), requires also the following set of agents:

1. **C**: a collection of control agents in charge of moving F over U, in order to cover the subspace V containing the objects o_i to be quantified, or

$$V \subseteq \{F_t\}_{t=1 \text{ to } u} \tag{7.1}$$

where F_t is the subspace of U sensed by S at the step t, and u is the number of steps required to cover V;

2. **G**: a collection of agents receiving information about the quantities δ of o_i identified by each R,

3. **A**: a collection of agents in charge of accumulating (σ) the quantities δ of o_i identified at each covering step t (e.g., Meck and Church, 1983) such that

$$\sigma_{t+1} = \sigma_t - \delta*\beta(1 - \omega/t), \omega > 1 \tag{7.2}$$

4. **Q**: a collection of agents classifying (υ_i) the quantities accumulated by A such that

$$\upsilon_i = d_i \text{ if } \alpha < \sigma < \beta, \text{ otherwise } \upsilon_i = \eta \tag{7.3}$$

where υ_i become the actual inputs to the quantifier $q_i \in Q$; $\alpha, \beta > 0$, and d_i is a label in the dictionary

$$D = [d_1, \ldots, d_n, \eta] \tag{7.4}$$

obeying the following rewriting rules $\oplus, \otimes$ used to order D:

$$d_{i+1} = d_i \oplus d_1, d_1 = d_1 \oplus \eta \text{ and } d_{i-1} = d_i \otimes d_1, d_1 = d_1 \otimes \eta \tag{7.5}$$

Let the classifying capability of Q to be dependent on the topographical location g of $q_g \in Q$ recognizing σ as υ_g (Fig. 7.1), that is

$$\upsilon_g = d_g \text{ if } \sigma - \alpha_g < \sigma < \sigma + \beta_g, \text{ otherwise } \upsilon_g = \eta, \tag{7.6}$$

$$\alpha_g, \beta_g = f(d_g) \quad (7.7)$$

If the possibility for $\mu_{di}(\sigma)$ of σ to be recognized by d_i is defined as

$$\mu_{di}(\sigma) \to 1 \text{ if } \sigma \to d_i \text{ or } \mu_{di}(\sigma) \to 0, \text{ if } \sigma \to \sigma - \alpha_g \text{ or } \sigma \to \sigma + \beta_g \quad (7.8)$$

then it may be proposed that the output n_g of the quantifier q_g is

$$n_g = \mu_{dg}(\sigma) \quad (7.9)$$

KFN works as follows. The controlling agents ($c \in C$) move the eyes to the locations where the objects to be counted reside. At each of these locations, the sensory neurons ($s \in S$) collect information to be used by agents ($r \in R$) in identifying the objects to be counted. Data concerning the number of objects recognized by R at each location is then delivered to the gating agents ($g \in G$). Each $g \in G$ may receive information from different sets of R, and from the same or different sensory systems. In this way, these agents free counting from sensory boundaries. Another task of the gating agents allows for block counting. Whenever more than one object may be jointly identified in a focused location (as, for example, in subtizing–Butterworth, 1999; Dehaene, 1997; Fink et al. 2001; Jensen et al. 1950; Piazza et al. 2002), their quantities may be the signal recorded by $g \in G$. When all objects in the focused location are recognized, and their quantities stored in $g \in G$, the controlling agents then move the eyes to the next location to be explored and signal the neurons $g \in G$ to send their data to the accumulators $a \in A$. The output of these accumulators is classified by the neurons $q \in Q$. The result of the counting is provided by those $q_g \in Q$ for which n_g is a maximum.

The KFN output n_g of q_g signals the confidence of σ being encoded by q_g as d_g. *Fuzzy Logic* and *Fuzzy Number* theories were introduced by Zadeh (1965) as the most adequate mathematical formalism for dealing with the types of approximate reasoning and calculation performed by humans. Fuzzy numbers are numbers of the type *around n* used to do most of our daily life quantification (Pedrycz and Gomide, 1998). KFN is a particular type of Fuzzy Number, and was specially tailored to describe the properties of counting in animals (Massad and Rocha, 2002, Rocha and Massad, 2003a). The above KFN circuit describes the main properties of those quantifying neurons, recorded at both the frontal and parietal lobes (Dehaene, 2002, Nieder et al. 2002, Nieder and Miller, 2003, Sawamura et al. 2002). Also, the parietal KFN circuit is proposed to correspond to the set of neurons forming the mental *number line* in humans (Butterworth, 1999; Dehaene, 1999; Fink et al. 2001; Gobel et al. 2001, Nuerk et al. 2001, Zorzi et al. 2002).

7.2.2 Quantification trajectory control

One of the most complex tasks in counting consists in the control of the counting pathway, because it is necessary to avoid counting the same element twice (or more), and to avoid forgetting an object in the counting collection. This type of control is part of the problem known as the "travelling salesman problem," a very well known one for AI scientists. It seems that animals (Brannon and Terrace, 1998; Carpenter et al. 1999) and children of school age learn to easily solve this task even for collections having a large number of objects. Children learn not to miss or miscount an object in the counting collection by labeling each counted element with a defined finger. Each identified object is associated to one (overt or covert) motor action (e.g., finger extension) by the children before they focus their eyes on the next element to be counted (Fig. 7.2). In this way, they easily discover when the same place is being visited twice, or when a possible unmarked object is found at the end of the counting trajectory.

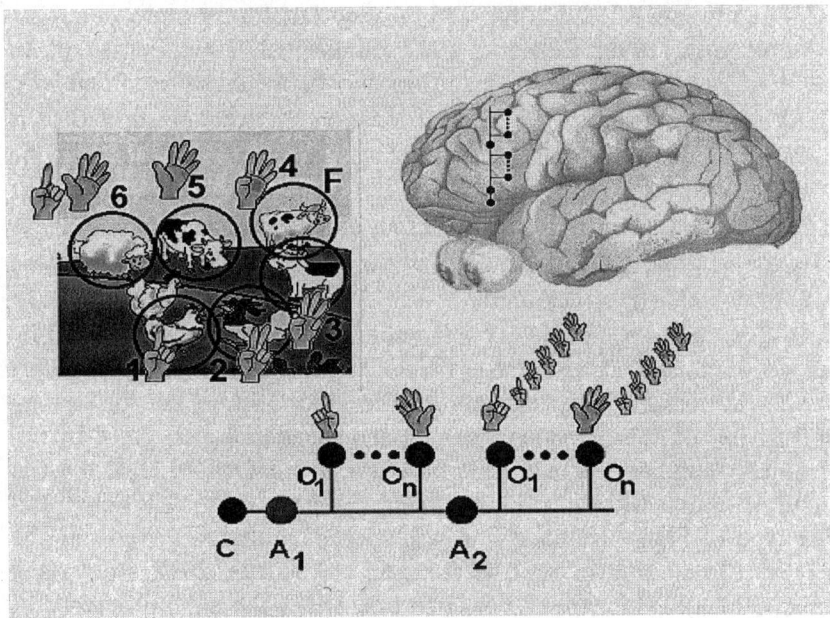

Fig. 7.2. Ordinal quantities: *Making sure everyone is counted; no one is missing.*

This kind of correspondence allows for the construction of another counting system—*the ordinal KFN numbers*—because each time the controller sends a signal to move the eyes to the next object, it may also send another pulse to an accumulator whose output is filtered by another class of neurons, called here *ordinal neurons* (Oi in Fig. 7.2), which in turn may be associated to any motor action. This may be the type of neurons recorded by Carpenter et al. (1999) in the monkey motor control cortex, which signals the order of a stimulus in a given sequence.

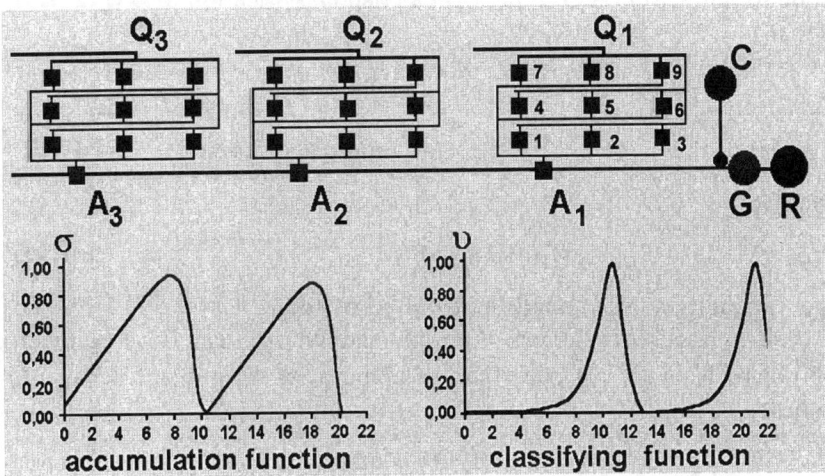

Fig. 7.3. Creating Base Fuzzy Numbers

7.3 The Crisp Numbers

The KFN circuit may evolve by increasing the complexity of A and Q (Fig. 7.3), that is, by:

1. changing from the linear accumulation (Fig. 7.1) defined by Eq. 7.2 to a non-linear function (Fig. 7.3):

$$\beta_{t+1} = Z_1 * \sigma_t * \beta_t^L * \delta, \quad \sigma_{t+1} = \sigma_t - Z_1 * Z_2 * \sigma_t * \beta_t^L * \delta + Z_3 \quad (7.10)$$

with **t** as the observation step, Z_1, Z_2 and Z_3 as constants;
2. creating a hierarchy of subsets of accumulators (A_1, A_2,...) to monitor the discontinuities on the lower level accumulators, such that whenever the accumulator A_i is reset, a pulse is accumulated by A_{i+1};
3. creating a hierarchy of sequential quantifiers (q_1, q_2,...) associated to that of accumulators (see Fig. 7.3), such that the accumulator A_i feeds the quantifier q_i.

7 Modeling of Arithmetic Reasoning

This KFN evolution creates a new kind of number system, called a Fuzzy Base Number System (FBN), whose base size is determined by the periodicity of the accumulating function. For instance, if the accumulator is reset after counting 10 objects, then the number base is 10, as in the decimal system.

KFN evolution may continue by adjusting the filtering or classifying function to allow $SQ_g \in Q$ to perform a crisp classification of σ (Fig. 7.3), such that

$$\upsilon_g = d_g \text{ if } d_g \otimes \gamma_g < \sigma < d_g \oplus \gamma_g, \text{ otherwise } \upsilon_g = 0 \qquad (7.11)$$

$$\mu_{dg}(\sigma) \approx 1 \text{ if } d_g \otimes \gamma_g < \sigma < d_g \oplus \gamma_g \text{ otherwise } \mu_{dg}(\sigma) = 0 \qquad (7.12)$$

because

$$\gamma_g = k(d_g) \to 0 \qquad (7.13)$$

In such a condition, SQ_g attends to A only if $\sigma = d_g$.

It is now possible to create a special family of Crisp Base Numbers (CBN) of KFN (Fig. 7.4), such that the output i_g of the g quantifier $i_g \in Q$ becomes

$$i_g = d_g \text{ if } d_g \otimes \gamma_g < \sigma < d_g \oplus \gamma_g, \text{ otherwise } i_g = 0 \qquad (7.14)$$

and the I output is made reentrant over the same set A of accumulators initially activating it, that is, if δ in Eq. 7.11 is made

$$\delta \leq i_g, g = 1, j \qquad (7.15)$$

In these conditions, the I output crisply codifies the cardinality of abstract sets bearing (or not) a relation to collections of objects in the real world. If the CBN base is set equal to 10, then we have the classical (or crisp) decimal number system.

The evolution of the KFN circuit provides animals with two counting systems, using fuzzy (Q) and crisp (I) quantifiers. Both systems may have different functions in sophisticated number systems like those developed by man. On the one hand, the Qs may easily control motor neurons in Broca's or the hand's areas to speak or write about the quantities identified by them (Fig. 7.4), because each Q may map to specific pre-motor and/or motor neurons. On the other hand, verbal sensory temporal neurons or visual parietal-occipital cells may directly map to I neurons, which in turn may change the adequate amount to the accumulators A_i (Fig. 7.4) to provide the adequate semantics for spoken and written numbers.

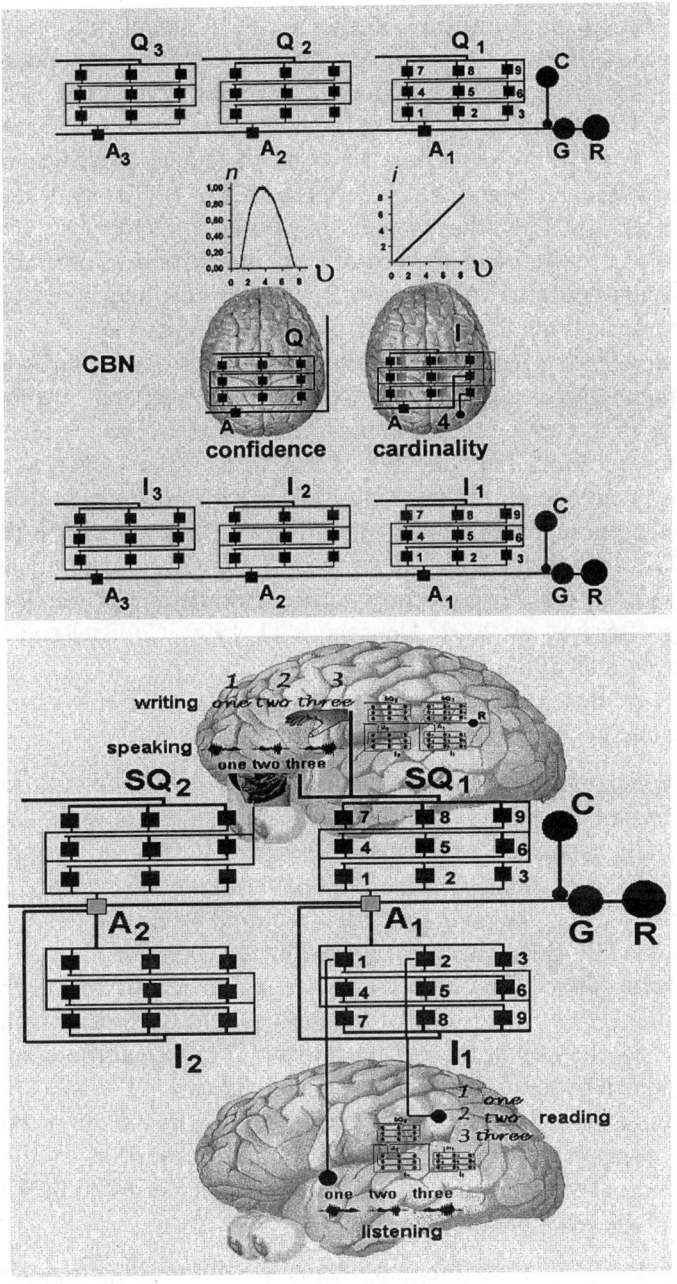

Fig. 7.4. Crisp numbers: *KFN evolution creates CBN number circuits*

There is a lot of redundancy in the complete model proposed here to cope with the processes of identifying quantities by animals and of counting by humans. This is a very important DIPS property. The same knowledge has to be available to different neighborhoods and to different agents specialized to solve the same task. Neurons that signal quantities are found both at parietal and frontal sites (Dehaene, 2002, Nieder et al. 2002, Sawamura et al. 2002) and they may be associated to the quantity identification (parietal neurons) and the use of this information in other neighborhoods (frontal neurons). Also, other frontal neurons may be in charge of computing ordinal numbers associated to sequences of motor actions (Carpenter et al, 1999), whereas parietal neurons assume the task of computing cardinal numbers associated to set *numerosity*. Ordinal and cardinal counting will agree in one to one counting when only one object is identified in the focused location, but they will differ in the case of block counting when more than one object are jointly identified (*subtizing*) in the focused location. The partial redundancy provided by KFN and CBN circuits is also in accordance with both experimental findings and theoretical propositions about the existence of approximate and crisp quantification by humans (e.g., Barth et al, 2003; Dehaene et al. 1999; Gallistel and Gelman, 2000; Nuerk et al. 2001; Polk et al. 2001).

Another interesting feature of the model is that a CBN may be a KFN evolution due to changes in the expression of:

1. genes which control the ionic channels in the membrane of the accumulator neuron; depending on the number and/or the quality of the Na, K, Ca and Cl channels, neurons may use monotonically increasing or periodic codes to encode the same kind of information (Rocha, 1992, Rocha, 1997). Thus, the ionic properties required by Eqs. 7.2 and 7.10 may be provided by changing ionic channel gene expression;
2. genes that control the number and quality of quantifying neurons; increase in cerebral size is one of the most salient features of animal evolution, and changes in the dynamics from Q to I quantifiers may easily be explained by ionic channel gene encoding; and
3. the complexity of the frontal circuits involved in the motor control of the counting trajectory; the growth and sophistication of the frontal processing capacity is one of the well documented differences between man and the other primates.

In consideration of the foregoing items, we propose that the genetic information for the development of CBN circuits was already available to the early hominids, and that increasing complexity in the lives of human societies may have pressed man to invent crisp arithmetic in at least four different cultures, viz., the Sumerian, Egyptian, Greek and Mayan cultures (Butterworth, 1999, Dehaene, 1997; Devlin, 2001, and Ifrah, 1985).

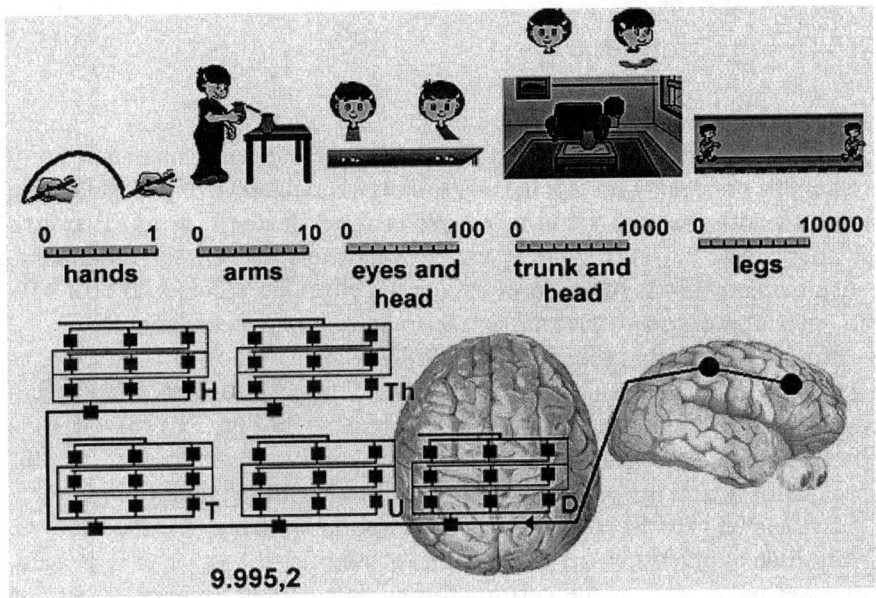

Fig. 7.5. The real numbers: *quantifying distances*.

7.4 The Real Numbers

The necessity for quantification extends beyond that of ordering actions and evaluating cardinalities. It also comprises quantification of distances. If the ability to quantify amounts of food and numbers of predators increases the odds of survival, so the ability to estimate distances in the different sensory-motor spaces is also a basic necessity for any animal, all of which need to quantify both cardinalities and continuous variables (Gallistel and Gelman, 2000). The KFN and CBN circuits are specially tailored for such a task. It suffices to substitute sensory input to the gate agents $g \in G$ with the output of the sensory-motor areas which determine the amount of movement in any of the different sensory-motor spaces (Fig. 7.5). For example, the different human action spaces—hands (very closed), arms (peri-personal), eyes and head (near), trunk and head (reachable) and legs (distant) spaces—all of these may be used by both KFN and CBN circuits to approximately or crisply evaluate distances and to support the genesis of the notion of real numbers (recently introduced in human culture) as well as the metric system, where it is used to measure distances.

7.5 Doing Arithmetic

KFN and CBN circuits may be designed for doing both fuzzy (Pedrycz and Gomide, 1998) and crisp arithmetic. On the one hand, man uses fuzzy arithmetic to deal with approximate daily calculations of the following type: spending around $**X** to buy more or fewer **Y** items; or to fill around **Z** liters of gas to travel more or less **W** kilometers, etc. On the other hand, crisp mathematics is required to maintain a checking account, to deal with complex commercial transactions, and to do scientific calculations.

Addition is neatly processed by both KFN and CBN circuits, whether as formal or simulated operations (Fig. 7.6). In the case of simulated operations, numbers representing quantities (chickens in Fig. 7.6) may first be decoded as a set of elements that are then used to mentally simulate the process of counting described in section 2.1. Each time a chicken is mentally imagined (or focused on), the corresponding value is loaded into G. Each time another chicken is mentally focused on, the value of G is accumulated by A, etc. In the case of formal calculations, spoken or written numbers are associated to the corresponding agents $i \in I$, which in turn load directly and adequately to the accumulators $a \in A$, such that the result is obtained by the classifications provided either by the neurons $q \in Q$ or i belonging to another set I. Hybrid calculations may be processed loading one number directly into $a \in A$ using the corresponding $i \in I$ and decoding the other as a set of elements, whose elements are sequentially accumulated in the same $a \in A$ through the gate $g \in G$. If the operands are ordered first, then this hybrid calculation may be optimized by loading directly the highest operand and decoding the others as a set of elements. Thus, both the KFN and CBN may allow man to make use of distinct adding strategies (Fayol, 1996; Gelmann and Gallistel, 1991; and Siegler, 1996). It is easily verified that simulated calculations may render addition time dependent on the size of operands, whereas formal calculations make it constant. Since different sets of agents may enroll to solve the same task proposed to a DIPS, then it may be that man activates different KFN and CBN circuits to a parallel processing of the same addition operation, the result being furnished on a "first come, first served" basis if speed is required, or by majority if precision is the goal. Also, the types of errors pointed by McCloskey, Harly and Sokol (1991) will be those expected for the kind of calculations proposed above, if mistakes are made in operand association

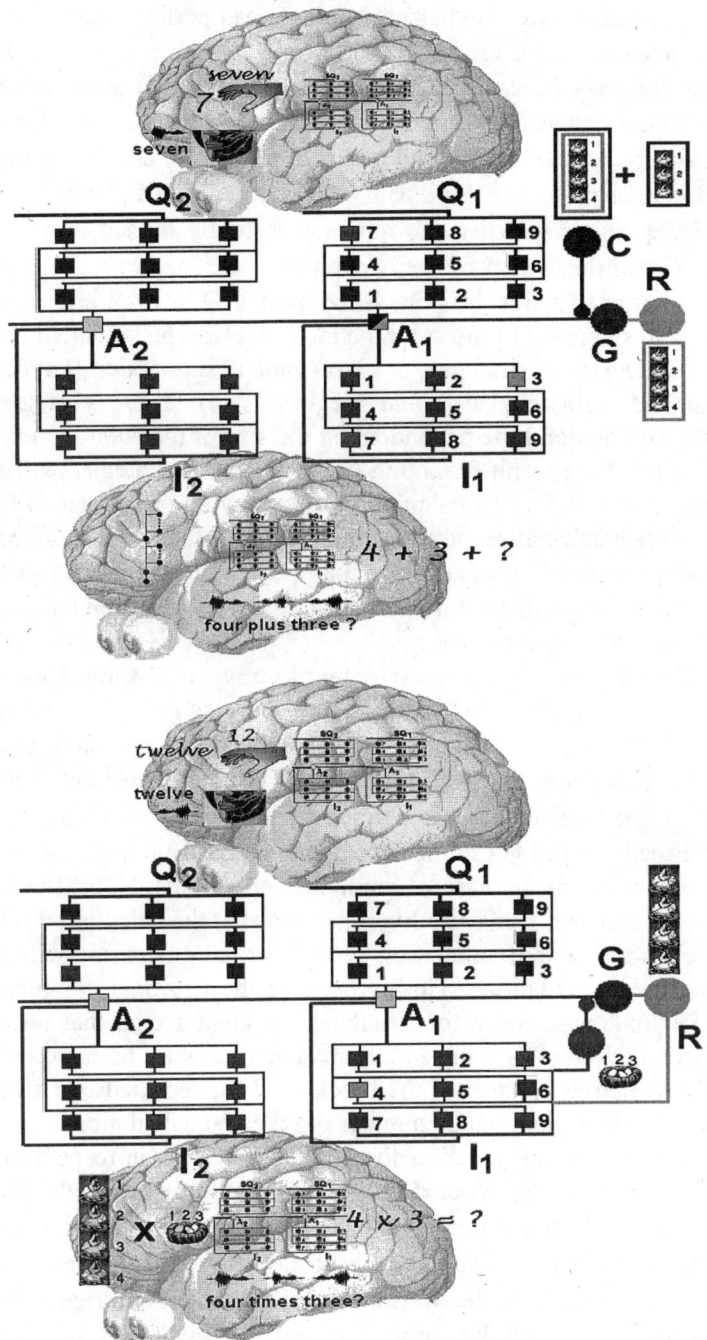

Fig. 7.6. Doing arithmetic: *adding and multiplying.*

of $i \in I$; in operand set decoding; by a $c \in C$ bad performance, or even by misclassifications of $q \in Q$ or $i \in I$.

Subtraction may be easily performed as the inverse operation of addition. In this case, the subtrahend is loaded in $a \in A$ and then $c \in C$ controls a step-by-step increase of the load in $a \in A$ until the classified output by $q \in Q$ or $i \in I$ equals the minuend. But subtraction may also be calculated by a decreasing counting, when the minuend is firstly loaded in $a \in A$ and then $c \in C$ controls a step-by-step decrease of the load in $a \in A$ using inhibitory neurons. The result is classified by $q \in Q$ or $i \in I$ after a number of steps corresponding to the subtrahend. These are the so-called *counting up* and *counting down* strategies used to simulate subtraction (Fayol, 1996; Gelmann and Gallistel, 1991; and Siegler, 1996). These strategies may render the subtraction time dependent on the size of the subtrahend (counting down) or of the result (counting up). Since both strategies will always be activated in a DIPS, the calculation time may be optimized, if the first circuit to finish calculation provides the solution. But subtraction may also be formally achieved by associating the operands to their corresponding $i \in I$, which in turn will load $a \in A$ through excitatory (minuend) and inhibitory (subtrahend) synapses.

Multiplication may be computed either as a repeated summation or as a block counting (counting by multiples). In the case of both simulated and formal calculations, one of the operands is loaded in $g \in G$ and repeatedly added to $a \in A$ under the control of $c \in C$. The difference between simulated and formal calculations is defined by the way $g \in G$ and $c \in C$ are loaded: directly from $i \in I$ in the case of formal computations, or by means of a set decoding in the case of simulated operations. Hybrid operations will be also allowed. It may be hypothesized that the calculation time will be greater in the case of simulation even if formal processing will also require repeated calculations. This is because the neuronal complexity required for loading g and c for simulation is greater than that needed for formal calculations. Also, the multiplication time may be expected to be inversely related to the size of the block used for repeated operations and to be directly dependent on the number of repeated calculations.

Division may be calculated as the inverse operation in respect to multiplication or as a case of repeated subtraction. In the first case, the divisor is loaded in $g \in G$ and repeatedly added under the control of $c \in C$ until the output at $q \in Q$ or $i \in I$ equals the dividend. In the second case, the dividend is loaded in $a \in A$; the divisor loaded in $g \in G$, and repeatedly subtracted from $a \in A$ until the output $q \in Q$ or $i \in I$ equals 0. In both cases, the result is the number of repeated pulses of $c \in C$ computed by the frontal ordinal circuit.

As in the case of multiplication, the repeated calculations for division may be either simulated or formally computed. By the same reasoning discussed for multiplication, the time for simulation will be greater than that for formal computations. In the case of simulation, the division time may become directly dependent on the size of the divisor or inversely proportional to the size of the dividend. But division may also be simulated as sectioning of the number line fed by sensori-motor data and generating real numbers (Zorzi et al, 2002). This may be a simulation that is independent of the size of operands.

7.6 The Evolution of Arithmetic Knowledge

The model proposed here is based on the assumption that initially a set of DIPS units (neurons) are able to discriminate the objects of interest in the real world and to provide basic information about their quantities to other specialized agents in charge of accumulating the results of sequential ordered observations. The accumulator output is then classified by another set of units, called here *quantifiers,* into a small set of classes allowing a primitive manipulation of quantities.

The *quantity of* sense was introduced here by associating a sensory *identifier* to two other different agents: the *accumulator* and the *quantifier* neurons, such that the number of identified objects in a given sensory system at each focused location is accumulated, followed by an estimation of their quantity. However, in such a simple sensory-specific model, let us say, five pictures have no relation to five sounds, even when both are associated to the same set of real world objects. This is because each kind of sensory information is specifically channeled to a defined set of accumulators, which in turns activates a specific set of quantifiers within a given sensory modality. In this way, we may speak of different quantifying circuits, each one sensory bounded.

The first important evolutionary step from the *quantity of* concept to that of *numerosity* was to channel different sensory information to the same set of gates, which in turn feed the same accumulators and quantifiers. Now, five pictures and five sounds are associated to the same quantity and may now refer, if necessary, to the same set of real world objects. This is because the same quantifying circuit is activated by information provided by different sensory systems. This was what may have happened in evolution with the emergence of the multi-sensorial (parietal) cortex.

The second important step in the development of the concept of *number* required the creation of a specific set of quantifiers—the I quantifiers—and the reentrance of the output from the I quantifier to the same gates G feeding the accumulators A, providing information to the very same I quantifiers. This loop freed quantification from any outside sensory input. In other words, reentrance freed quantification from observation in the external environment. Therefore, this step created the concept of *number* as a set of symbols referring to quantities, or the cardinality of abstract sets. This generated number lines in the parietal cortex, as proposed by many authors. (Butterworth, 2000; Dehaene, 1999; Fink et al. 2001; Gobel et al. 2001, Zorzi et al. 2002). It is worth stressing that reentrance is an important feature of brain processing, as mentioned by Edelman (1989 and 1997) and Rocha et al. (2001). This highly evolved model supports the human number capability and perhaps a degree of *number sense* in the other most evolved primates.

Both the accumulation and filtering functions used to generate any of the above evolutionary steps of our model may be understood as supported by ionic properties of the neural membrane. Different dynamic membrane systems may generate monotonically increasing or periodic changes in the membrane potential (Rocha 1992; Rocha 1997). Different genes governing the expression of ionic channels may be responsible for such different encodings and changes in the expression of these genes may account for KFN evolution. These ionic genes are known to have appeared early in brain phylogeny. This implies that at least the concept of *numerosity* may be at hand for many species. The distinct spatial configurations of the DIPS agents required by the different steps of our model may also be easily understood from our knowledge of brain phylogeny, and governed by genes responsible for axonal growth and addressing (e.g., Edelman, 1997). Again, these genes are known to be ancient.

The *number sense* is now a well-known animal capability (Butterworth, 1999; Dehaene, 1999; Devlin, 2001) and humans invented numbers at least in four different cultures at four different times (Ifrah, 1985). In this context, we may propose that an *Animal Mathematical World* is a product of the interaction between genes and their habitats, and by the same token the *Human Mathematical World* is determined by the interplay between the brain (genes) and culture (habitat).

7.7 Representation of Other Abstract Entities

Mathematics is much more than the 'science of numbers.' If mathematics is considered as the *science of patterns,* we can demonstrate that other abstract mathematical constructs may be modeled in a similar way to the manner in which the number concept was built up, step-by-step, towards increased generalization. Patterns are stable relations among features of concrete or abstract objects recognized by a set R of specialized agents.

The same general procedure used above to construct both KFN and CBN numbers may be applied to construct other abstract mathematical concepts, starting from the identification of other specific patterns in the environment. In the same way that identifiers provided accumulators with information about quantity, these operators may furnish other types of information about defined sensory patterns (e.g., line relations) to other specific processing agents (e.g., angle evaluation), so that similar kinds of (geometric) concept generalization are achieved through multi-sensory pattern channeling, with information reentrance through other specialized processing circuits. As a matter of fact, there is no *a priori* difference between visual pattern recognition and classical geometric processing. What is the difference between the recognition of the pattern of an elderly male human face by some specialized cell at the temporal lobe and the formal construction of an object in any equivalent order-dimensional geometric space? Certainly the increased brain capacity for geometric analysis of visual information in some animals is also a characteristic with high adaptational value.

How is it possible to think about something that does not exist? This is a high level of abstraction (level 3 of abstraction according to Devlin, 2001) that may be supported by simulation performed by DIPS agents using abstract objects of the type described above for integers. Simulation only requires other specialized agents capable of recruiting abstract level 3 agents. DIPS simulation is the keystone of imagination and it is supposed to appear in evolution with the enhancement of the frontal lobe, especially in humans. In fact, imagination is nothing but mental simulation controlled by a top-down process, where most frontal neurons recruit more posterior agents in an orderly process. In such a way, DIPS simulation may allow man to create a sophisticated number theory or other abstract mathematical object.

8 Brain Maps of Arithmetic Processes in Children and Adults

Despite the increasing number of experimental mappings of a widely distributed neural circuit subserving human arithmetic cognition, the theoretical reasoning concerning these data remains mostly metaphorical, and guided by a connectionist approach. Herein, the **DIPS** theory developed in previous chapters is used to develop a new technique for EEG mapping of the brain activity associated with cognition, and to study arithmetic cognition in elementary school aged children and in adults. Performance in solving arithmetic calculations was studied in adults and children while their EEG activity was recorded. Experimental data showed a) a clear-cut distinction between genders, males being faster than females in providing equally correct answers; b) quick learning, characterized by a decrease in calculation time, which is dependent on the order of problem presentation; and c) a number size dependence that differed between children and adults. Factor analysis showed three distinct patterns of neuronal recruitment for arithmetic calculations in all experimental groups, which varied according to the type of calculation, age, and gender.

8.1 The Study of the Mathematical Brain

Experiments carried out in humans demonstrated the involvement of the inferior parietal cortex as well as multiple regions of the prefrontal cortex in the course of numerical performance (e.g., Butterworth, 1999, Cabeza and Nyberg, 1997, 2000; Carpenter et al., 1999; Dehaene, 1997; Dehane et al. 1999; Göbel et al. 2001; Iguchi and Hashimoto, 2000; Jahanshahi et al. 2000; Kong et al. 1999; Menon et al. 2000; Nieder et al. 2002; Pesenti et al. 2000; Ratinck et al. 2001; Sawamura et al. 2002; Skrandies et al., 1999; Zorzi et al. 2002). It is now accepted that the inferior parietal region is important for the translation of numerical symbols into quantities, and the representation of relative magnitudes of numbers. The prefrontal cortex is proposed to be responsible for the processes involved in sequential ordering of successive operations, in control over their execution, error correction, inhibition of verbal responses, etc. But other central and temporal cortical areas are also involved in many arithmetic calculations. It seems, therefore, that human arithmetic capabilities result from complex cerebral processing involving different types of neurons widely distributed throughout the brain, each of them in charge of solving a particular subtask of the whole problem.

Despite the increasing number of authors (e.g., Cowell et al. 2000; Dehaene et al. 1999; Jahanshahi et al. 2000; Zago et al. 2001) experimentally mapping a widely spread neural circuit supporting human arithmetic cognition, the theoretical reasoning about these data is not well developed. Although neurons at distinct areas in the brain are assumed to take charge of different duties in the solution of the experimental task, the results are always discussed by hypothesizing some association between the different areas without questioning any distinct behavior at the level of the neurons at each of these areas. Also, the role played by each brain area in the different arithmetic tasks is, in general, discussed metaphorically rather than formally. In the **DIPS** approach, however, the task of counting is assumed to be performed by a large number of specialized agents distributed all over the brain, such that mathematical skills becomes dependent on both the different types of specialization by different agents as well as on how these agents engage themselves in solving given mathematical problems.

Another interesting **DIPS** property is that many different means are tried to solve the very same arithmetic problem. Indeed this seems to be the strategy used by the brain in the case of arithmetic processing. Many studies of how children in different countries solve standard arithmetic problems has revealed that the subjects use multiple different strategies such as (Butterworth, 1999; Dehaene, 1991; Dehane, 1997; McCloskey et al., 1991; Siegler, 1996):

1. **total manipulation**: the child counts each set to be processed separately by pointing, marking, etc. each of their elements, and then counts by the same process each element of the union, or the complement, etc. of these sets to get the final result;
2. **simplified manipulation**: the result is obtained by counting each element of the union, or the complement, etc. of the sets to be processed;
3. **optimized manipulation**: the result seems to be obtained by performing the minimum counting, which varies according to the type of calculation to be performed, and
4. **mental calculation**: the results are quickly processed by specialized circuits. Also, the type of function explaining the number size effect reported in the literature correlates with the kind of manipulation used (e.g., McCloskey et al, 1991).

Finally, training changes the frequency of use of the different types of strategy, such that adults tend to rely more on mental calculation than children (Siegler, 1996). Rocha et al., (2003c,d) showed that children and adults use different strategies for solving any kind of arithmetic calculation, because they found different types of correlation between the calculation time and the size of the different operands. Also, they showed that the size effect dependence was more complex for adults in comparison with children, and they concluded that learning enriches arithmetic knowledge by increasing the number of strategies available for the same calculations.

Here, the distributed properties of arithmetic cognition were studied in three experimental groups while having their EEGs recorded in the course of solving arithmetic calculations (Rocha et al., 2003c). The performance in solving the arithmetic problems was measured by the time interval required for the calculation, since the error rate was low in all groups. The EEG analysis provided different brain mappings (Rocha et al., 2003c) that revealed the distributed properties of the mathematical brain.

8.2 The Technique

The DIPS approach to cerebral physiology also results into new technologies for the analysis of EEG activity related to cognitive tasks (Foz et al. 2001; Rocha and Rocha, 2002; Rocha et al., 2003c).

8.2.1 Theory

Definition 8.1: Let $\rho(n_i, n_k)$ measure, in the closed interval $(0,1)$, the possibility of any two neurons n_i, n_k to jointly involve themselves in solving a task t, such that

$$< \rho(n_j) > = \frac{1}{n} \sum_{k=1}^{n} \rho(n_j, n_k) \quad (8.1)$$

$$< \rho(n_j) > -0.5 = \xi \quad (8.2)$$

N is said to be a *strongly (un)connected* system N_s if $\xi \to \pm 0.5$ for most of $a_j \in N$. Strongly (un)connected systems are of no interest where cognition is concerned, because either their agents have difficulties in enrolling together to solve a task, or the relations shared by their agents tend to be stereotypical rather than versatile. Also, their agents are more likely to share strong, rather than plastic, commitments.

N is said to be a loosely connected system N_L if $\xi \to 0$ for most of $a_j \varepsilon\, N$ even if $\rho(a_j, a_k) \to 1$ or 0 for some, but not all $a_k\, \varepsilon\, N_L$. A DIPS is a loosely connected system, because each of the agents retains the maximum capability to enroll with different groups of other agents in the effort to solve different tasks, since $\rho_m(n_j) \cong .5$.

Definition 8.2. The following brain entropies are defined from Eqs. 2.2 to 2.25:

1. **the commitment $h(n_{j,k})$ of n_j, n_k** in jointly solving a given task is calculated as:

$$h(n_{j,k}) = -\rho(n_j, n_k) \log_2 \rho(n_j, n_k) - {\sim}\rho(n_j, n_k) \log_2 {\sim}\rho(n_j, n_k) \quad (8.3)$$

$$\sim\rho(a_j, a_k) = 1 - \rho(a_j, a_k) \quad (8.4)$$

2. **the enrollment capability $h_m(n_j)$ of n_j** is calculated as

$$h_m(n_j) = -\rho_m(n_j) \log_2 \rho_m(n_j) - {\sim}\rho_m(n_j) \log_2 {\sim}\rho_m(n_j) \quad (8.5)$$

$$\sim\rho_m(n_j) = 1 - \rho_m(n_j) \quad (8.6)$$

that is, $h_m(n_j)$ is a function of the mean probability $\rho_m(n_j)$ of n_j to communicate with the other agents $n_k\, \varepsilon\, N$, and

3. **the actual commitment $h(n_j)$** of n_j to solve the task is

$$h(n) = \sum_{k=1}^{n} h_m(n_j) - h(n_{j,k}) \quad (8.7)$$

Proposal 8.1. The linear correlation coefficient $r_{i,k}$ for the EEG activities recorded at site d_i, d_k is assumed to be an indirect measure of the possibility $\rho(n_i, n_k)$ of the enrollment of the neurons n_i, n_k at these locations in solving a given task. In this way, the (mean) EEG activity recorded, and associated to a given event **e** of such task **t**, is used to calculated the corresponding $h_e(n_j)$ for each recording site d_i. The set $\mathbf{H_e}$ of all $h_e(n_j)$ values for all similar events **e** are used to generate different brain mappings (fig. 8.2) according to the statistics applied to analyze H_e.

Remark 8.1. Foz et al. (2001) used this technique to study the plasticity of neural circuits for language in brain-damaged children, where the classical Broca or Wernicke areas were destroyed during their fetal life. These children experienced severe delays in language development, but by the age of 11 years their brain mappings, associated with oral charades and texts, revealed an extensive use of the right hemisphere areas to compensate for the lesion of their homologous left sites.

8.2.2 The Methods

Data were obtained from the three experimental groups by Rocha, Rocha and Massad (2003 c,d). These groups were equal in regard to gender balance, in mean age differences, and in cognitive development.

1. Group **CHI2**: 20 children — mean age: 7 yrs, 7 mo; enrolled in 2nd and 3rd semesters of elementary school; mastering addition and subtraction, as well as reading and writing simple phrases;
2. Group **CHI4**: 24 children — mean age 8 yrs, 3 mo; enrolled in 3rd and 4th semesters of elementary school; mastering addition, subtraction, and multiplication; in their initial training for division; also, reading simple texts and writing simple phrases;
3. Group **AD**: adults: 20 adults — mean age 28 yrs; enrolled in graduate courses in the field of exact sciences in a university near São Paulo, and attending a special training program in Biotechnology.

The children were selected from a group whose parents agreed with the explained experimental protocol and who signed a special permission form. Each group was formed by children from two equivalent classrooms in respect of their cognitive profile, as evaluated by the principal of a middle class school in the city of Guarulhos. Adults volunteered after having the experimental protocol explained to them and they also signed the permission form, whose terms were approved by the university's Ethics Committee.

The experimental protocol consisted in solving 30 different problems (Fig. 8.1) for each arithmetic calculation while the volunteers' EEGs were recorded (Fig. 8.2). Each question was visually presented on a computer screen, and the volunteer had to choose the correct response among a set of displayed numbers. The questions were presented in two different visual formats. In one of them (**VA** format) the quantities were represented by numbers and elements of a given class of objects (toys, fruits, etc.), whereas in the other format (**VB** format) quantities were only represented by numbers. Two different series of questions were prepared, each concerning a given display format, and containing 15 problems each. The questions in each series were randomly selected when they were initially programmed.

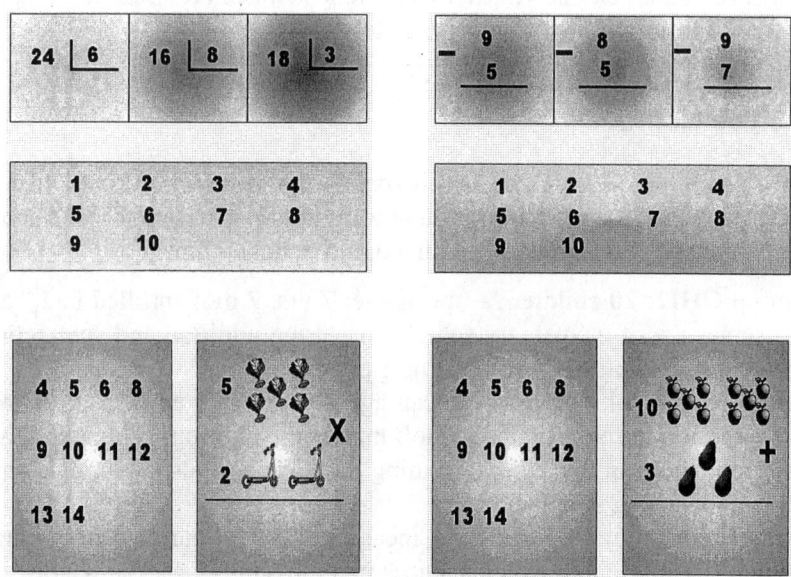

Fig. 8.1. The experimental protocol: *Testing arithmetic knowledge.*

Each question in the series was numbered to allow the study of any dependence of performance concerning the order of question presentation. Whereas adults solved both series, the groups **CHI2** and **CHI4** manipulated only the questions in the **VA** format. This was mainly because initial tests revealed that children could become tired if submitted to long series of calculations. **CHI4** children were tested in two different epochs: at the end of the 3rd and 4th semesters, in an attempt to quantify possible learning via reduction in errors and in response time. Adults solved the **VA** series for all four types of arithmetic calculations before the **VB** series were presented. This was done in order to allow comparisons between children and adults to be studied independently from the type of visual presentation, and to assure that the **VA** format could not be interpreted as infantile by the adults. All mistakes and correct responses were clearly signaled to the volunteers after they clicked their choice of answer to the question posed. The set of numbers available for response selection were displayed in lines of increasing but not consecutive quantities. These sequences were randomly selected when programming the experimental protocol, such that all volunteers were presented with the same response set for each question posed. All sessions were videotaped for further inspection of the volunteer performance whenever necessary. Adults were encouraged to comment on the experiment at the end of the session.

Fig. 8.2. Game playing while 20-electrode EEG is registered

Each subject solved the tests while his/her EEG was registered with 20 electrodes placed according to the 10/20 system; impedance smaller than 10 K ohm; low band passing filter 50Hz; sampling rate of 256 Hz and 10 bit resolution (Fig. 8.3).

Two networked personal computer were used: one for the EEG recording and the other for game playing. Timing of test events (e_1, e_2, ...), like the beginning of the visual information display, decision making, etc., were written as corresponding marks (m_1, m_2, ...) in the file of the recorded EEGThe EEG was visually inspected for artifacts before processing, and the events associated with a bad EEG were discarded. The mean time for solving the text was calculated for each experimental group. The recorded EEG was averaged for epochs of duration equal to this mean time, referred to the beginning of the visual display, in order to generate the Game Event Related Activity (**GERA**) file for each volunteer (Fig. 8.4).

160   8 Brain Maps of Arithmetic Processes in Children and Adults

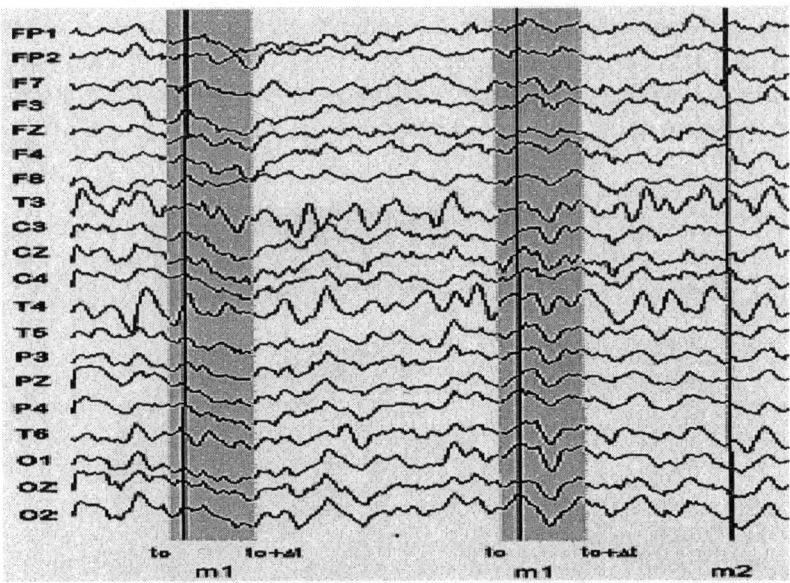

Fig. 8.3. EEG Recording

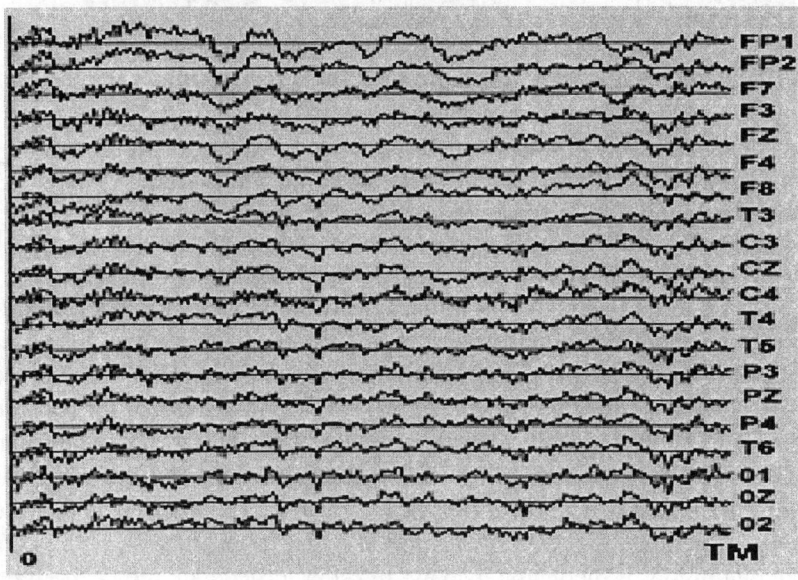

Fig. 8.4. Game Event Related Activity

These **GERA**s were used to calculate different brain mappings according to the following rationale (Foz et al. 2001; Rocha et al. 2003c). The linear correlation coefficients $r_{i,j}$ for the averaged activity at each recording site d_i, referred to the averaged activity for each of the other 19 derivations d_j, were calculated for each GERA (Fig. 8.5).

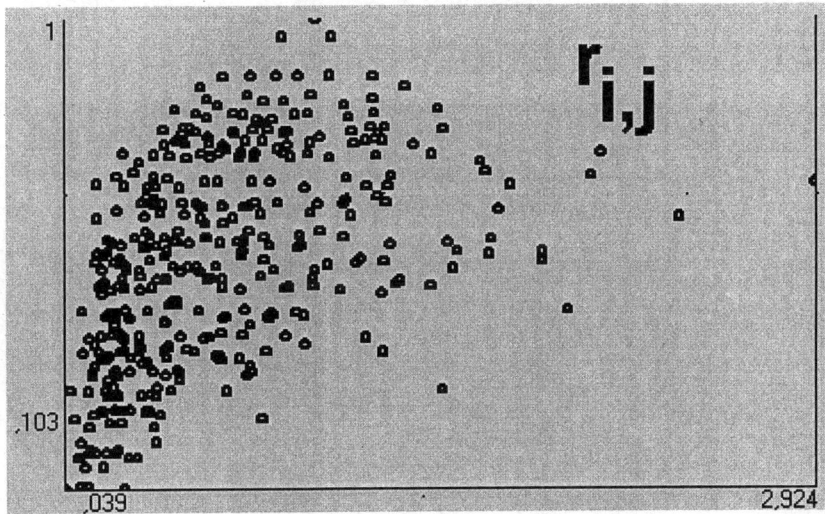

Fig. 8.5. Regression Analysis

This $r_{i,j}$ is considered a measure of the possibility $\rho(n_i, n_j)$ that neurons n_i at d_i are exchanging information with those neurons n_j at d_j, and this number was used to calculate $h(n_j)$ for each d_i (Fig. 8.6). These data were used to calculate the following brain mappings: a) **MCC**: plot of the normalized mean $h(n_j)$, and b) **FM**: three factors were extracted by using Principal Components Analysis, and these factors were rotated using the varimax normalized method. If these extracted factors explained more than 50% percent of the total $h(n_j)$ variability, then the analysis was considered acceptable, and the color plots of the calculated correlation coefficient for each electrode and each factor were produced.

The reaction (or calculation) time was measured by the computer as the time interval from the beginning of the visual display up to the moment the volunteer clicked a number as the selected answer. Whenever an error occurred, the volunteer was signaled for error and the computer clock reset. The computer also kept track of the number of errors and correct responses of each volunteer. Non-parametric static was used to evaluate differences between groups and gender, and multiple regression analysis was employed to study possible correlations between calculation time, number size, order in the calculation series and gender.

VOL	SEX	C3	C4	CZ	F3	F4	F7	F8	FP1	FP2	FZ	O1	O2	OZ	P3	P4	PZ	T3	T4	T5	T6
1	1	2,371	2,37	1,97	1,96	2,21	1,69	1,55	2,18	2,05	2,4	1,9	1,54	1,2	1,12	1,73	1,64	0,69	0,61	1,38	1,19
2	1	1,631	2,19	2,01	2,03	2,57	1,04	0,58	2,03	2,73	2,27	1	1,07	0,9	1,44	1,38	1,71	0,94	1,85	1,03	0,54
3	1	3,532	3,17	4,15	4,36	5,07	3,13	1,59	6,14	4,14	6,39	3,8	3,31	3,2	2,87	1,34	2,66	2,31	1,97	3,11	2,71
4	1	1,189	1,71	1,51	0,75	1,72	1,87	2,67	0,98	1,68	1,61	0,7	1,76	2	1,15	2,1	1,73	0,91	2,2	0,74	1,77
5	1	3,49	2,26	2,61	2,92	2,38	2,77	4,14	2,55	3	2,14	2,6	1,94	1,2	2,86	2,3	2,96	2,55	3,87	2,28	2,19
6	1	3,371	2,03	2,6	2,67	2,26	2,9	1,58	3,71	2,76	3,47	2,6	2,21	2,7	2,74	2,27	3,29	1,72	1,57	1,68	0,6
7	1	4,202	3,83	4,01	3,8	3,24	1,91	2,87	2,38	2,94	3,39	3,1	1,93	2,2	3,3	2,6	3,78	2,04	1,78	1,02	2,03
8	1	1,398	2,36	2,35	3,93	4,48	4,31	4,24	5,23	5,25	5,25	3,4	3,08	3,1	0,74	0,82	0,84	0,66	1,1	1,9	1,3
9	1	1,998	3,56	2,05	1,48	3,23	4,2	2,89	3,53	1,34	2,24	1,9	2,58	2,3	1,78	3,03	2,13	3,22	3,14	3,14	3,5
10	1	2,382	2,03	1,56	2,87	3,22	2,39	4,06	3,97	4,04	3,79	1,3	1,05	1,2	1,34	0,97	1,03	2,1	2,09	1,82	1,09
MEAN		2,556	2,55	2,48	2,68	3,04	2,62	2,62	3,27	2,99	3,29	2,2	2,04	2	1,93	1,85	2,18	1,71	2,02	1,81	1,69
11	2	2,726	1,89	2,75	3,08	1,95	2,34	2,93	3,94	3,93	4,1	1,5	1,93	1,3	1,37	2,37	1,3	1,32	2,99	1,22	1,23
12	2	0,662	0,57	0,65	0,83	0,63	1,03	0,5	0,86	0,6	0,74	0,1	0,32	0,3	0,39	0,4	0,49	0,6	0,51	0,24	0,32
13	2	2,234	1,85	2,64	2,5	2,14	3,18	3,4	2,9	3	2,16	2,3	2,29	2,6	1,67	2,57	1,29	2,5	2,42	1,81	2,42
14	2	0,111	0,13	0,15	0,22	0,15	0,37	0,24	0,3	0,11	0,14	0,1	0,22	0,2	0,13	0,25	0,11	0,28	0,18	0,16	0,15
15	2	2,757	2,03	1,51	2,56	2	3,58	3,97	2,77	1,86	1,26	1,2	1,09	0,9	2,48	1,22	0,87	2,72	1,93	2,66	1,27
16	2	2,238	2,38	1,6	2,35	2,62	2,56	3,94	2,61	3,81	3,05	1,3	2,23	1,8	2,16	0,87	0,69	1,58	2,89	1,52	2,03
17	2	1,299	2,32	2,08	2,7	3,83	2,02	2,28	4,29	4,12	3,91	1,9	2	2,4	0,53	0,91	0,62	0,83	1,14	2,21	0,35
18	2	1,023	0,89	1,61	1,46	1,86	1,23	0,97	1,69	2,38	2,5	1,6	1,23	1,8	1,37	1,13	0,71	0,83	1,38	1,05	1,57
19	2	2,065	1,65	1,79	2,73	2,66	3,35	2,46	2,91	2,14	2,79	1	1,12	1	1,65	0,75	1,32	2,08	0,47	1,48	1,26
20	2	0,627	0,9	0,67	0,87	0,91	0,85	1,21	0,96	1,04	0,85	0,8	0,54	1	0,94	0,54	0,73	0,8	0,46	0,67	0,4
MEAN		1,574	1,46	1,55	1,93	1,88	2,05	2,19	2,32	2,3	2,15	1,2	1,3	1,3	1,27	1,1	0,81	1,35	1,44	1,3	1,1

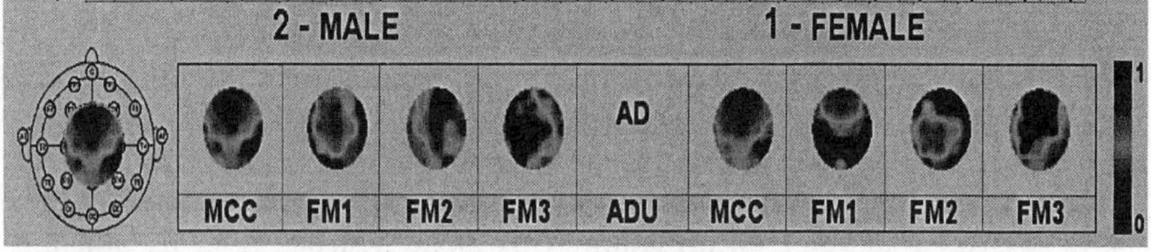

2 - MALE   1 - FEMALE

| | MCC | FM1 | FM2 | FM3 | AD ADU | MCC | FM1 | FM2 | FM3 |

Fig. 8.6. EEG Correlation Analysis

8.3 Agent Commitment Experimentally Measured

The mean, maximum and minimum values of the actual commitment ($h(d_j)$) differed among groups and type of calculations (See Table 8.1). The differences between the mean values obtained for the adults and children were statistically different for all type of calculations, as well as for male and females. No statistical difference was observed between CHI2 and CHI4 groups.

Table 8.1. Commitment ($h(d_i)$) vs. Calculation

	ADU						
	MALE			FEMALE			
	MEDIA	MAX	MIN	MEDIA	MAX	MIN	
AU	2,12	9,01	0,07	2,50	8,26	0,11	*
SU	2,08	7,04	0,13	2,42	7,62	0,27	*
MU	2,08	6,76	0,00	2,59	8,21	0,35	*
DI	2,31	6,66	0,24	2,67	7,62	0,48	*

	CHI4						
	MALE			FEMALE			
	MEDIA	MAX	MIN	MEDIA	MAX	MIN	
AU	3,18	11,24	0,12	3,29	9,34	0,00	*
SU	2,60	6,98	0,25	3,16	8,51	0,00	*
MU	2,47	6,38	0,04	3,41	9,48	0,00	*
DI	2,56	7,65	0,11	3,08	7,52	0,00	*

	CHI2						
	MALE			FEMALE			
	MEDIA	MAX	MIN	MEDIA	MAX	MIN	
AU	2,65	6,97	0,16	2,85	9,34	0,11	
SU	2,60	6,79	0,13	3,15	8,51	0,27	*

AU: Addition; SU: Subtraction; MU: Multiplication; DI: Division. ADU: adults; CHI4: children enrolled in 4th and 5th semesters of elementary school; CHI2: children enrolled in 2nd and 3rd semesters of elementary school. Mean, Max and Min: mean, maximum and minimum $h(d_i)$ values, respectively. Statistical differences between male and female means are marked with a *.

The commitment $h(d_j)$ of any agent d_i in solving a given arithmetic calculation, was calculated according to Eqs. 8.1 to 7, taking into consideration that the correlation coefficient $r_{i,j}$ between the EEG activity recorded at the sites d_i, d_j, measures the possibility $\rho(n_i, n_k)$ that the neurons n_i, n_j in these sites enroll themselves in the solution of the proposed task. The $h(d_j)$ max/min values in Table 8.1 clearly show the adequacy of this formal approach, since the actual values $h(d_j)$ calculated for each arithmetic calculation and each experimental group obeyed the theoretical conditions required to guarantee that agent commitment has always a positive value. Also, the mean $h(d_j)$ was statistically smaller for male than females in all groups, which may be an explanation of the fact that the males were quicker than females in solving arithmetic questions (Rocha et al., 2003c). Finally, the mean $h(d_j)$ was smaller for adults in comparison to children, but adults were faster than children in doing the same calculations.

8.3.1 Males are Faster than Females in Arithmetic Calculations

One of the main results of the present experiments was a clear gender difference for all groups. In general, males were quicker than females in arithmetic calculations (Fig. 8.7), but there was no consistent gender difference concerning errors.

Adult males were quicker than adult females in all kinds of arithmetic calculations, although the difference tended to be smaller in the case of multiplication. These differences were statistically significant at the level of $p < 5\%$, as evaluated by the Man-Whitney U test. Also, the index $I = (T_m - T_f)/SD$ is that proposed by Halpern (1992) to quantify gender differences was calculated for each arithmetic task. (SD is the standard deviation calculated for the entire group ADU, and T_m, T_f the mean male and female times, respectively.) This index was positive and greater for addition and division in comparison with subtraction and multiplication. It reached the highest value of 38% in the case of addition and the minimum value of 9% in the case of multiplication.

CHI4 males were quicker than their female classmates in the case of addition, subtraction and multiplication, but not in the case of division. Again, the differences were statistically significant at the level of $p < 5\%$ as evaluated by means of the Man-Whitney U test. The value of I steadily increased from 5%, in the case of addition, to 12% in the case of subtraction, and to 19% for multiplication. CHI2 females were slower ($p < 5\%$) than their male classmates in the case of addition and subtraction, but not in the case of multiplication. The value of I was 5% in the case of addition and only 1% for subtraction.

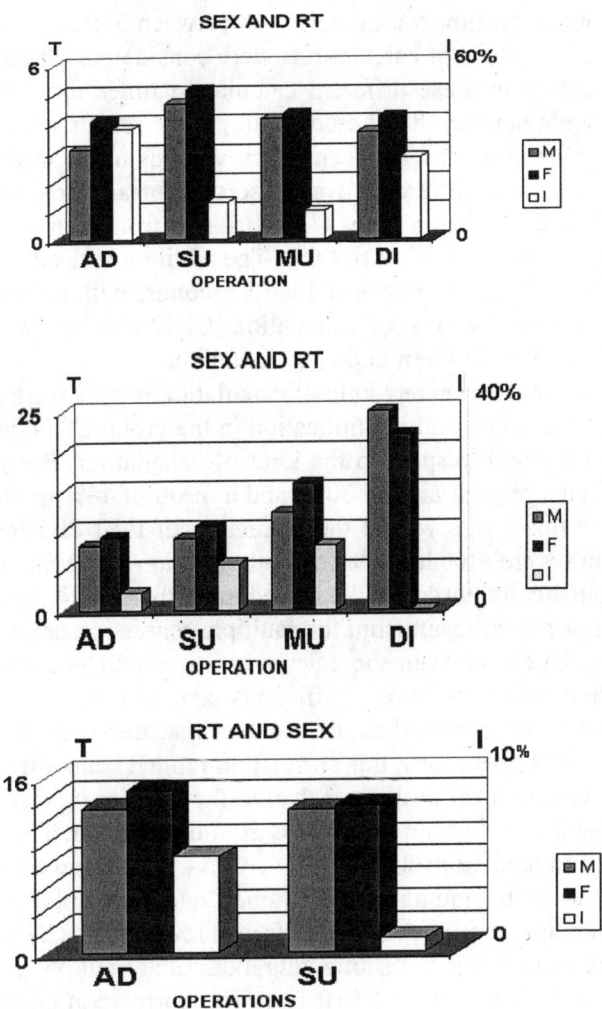

Fig. 8.7. Calculation time: *Males are faster than females in arithmetic calculation. Top: ADU; middle: CHI4 and bottom: CHI2*

Another general result was the difference in the calculation time between the experimental groups. Adults were quicker than children in all kinds of calculations, and CHI2 were slower than CHI4 children in adding and subtracting. All these differences were established at a level of statistical significance of $p < 5\%$, as evaluated by means of the Man-Whitney U test.

The calculation time varied narrowly between 3–5 seconds in the group ADU when all kinds of calculations were considered. There was no statistical difference in these different calculation times for this group. CHI4 children spent between 8–12 seconds to get the results in the case of addition and subtraction and between 12-16 seconds in the case of multiplication. No statistical difference was observed for addition and subtraction, but female multiplication times were different from their addition and subtraction times at the level of $p < 6\%$. The addition and subtraction times in the group CHI2 varied between 13–15 seconds, with no statistical difference concerning the type of calculation. CHI2 children were statistically slower than CHI4 children in their calculations.

Adults did not err on any kind of calculation. Errors were around 3% for addition, subtraction and multiplication in the group CHI4, and did not statistically vary with respect to the kind of calculation. However, the error rate reached a peak at around 70%, and a mean of 48% in the case of division, since this group was at the beginning of their division learning period. Errors were around 12% for both addition and subtraction in group CHI2. Statistically, these children erred more than CHI4 children.

The most robust result from the multiple regression analysis was the inverse correlation between the calculation time and the order of question presentation (see correlation coefficients labeled L in Fig. 8.8). The ADU calculation time decreased as the order of question presentation increased for all kinds of operations; this correlation ranged from 40% to 60%. CHI4 children were quicker at the end than at the beginning of the addition and subtraction series, but not in the case of multiplication. The correlation for this group ranged from 21% to 36%. CHI4 children were quicker at the end than at the beginning of the addition and subtraction series, too. But their correlation coefficients ranged from 11% to 30%. CHI4 children were also quicker at adding and subtracting, but not at multiplication, in the second test as compared to the first one. The correlation coefficients in this case varied between 13% and 18% (see correlation labeled Epoch in Fig. 8.8).

The calculation time was also gender related for all kinds of calculations in the case of the ADU group; for addition and subtraction in the case of group CHI4; and for addition in the case of CHI2 children. The correlation was around 20%, and showed male calculation time was shorter than female processing time.

"Sex differences in mathematics achievement are well documented", as Halpern put it in her 1992 book: *Sex differences in cognitive abilities*. The importance of this subject is clearly acknowledged (Beal, 1999) in the special issue of the journal *Contemporary Educational Psychology* devoted to the Math-Fact Retrieval Hypothesis, proposed by Royer and Tronsky

(1999) to explain that males outperformed females in the SAT-Math exam because they are quicker in math-fact retrieval. Finally, Gallager et al. (2000) proposed that "strategy flexibility is a source of gender differences in mathematical ability assessed by SAT-M and GRE-Q problem solving."

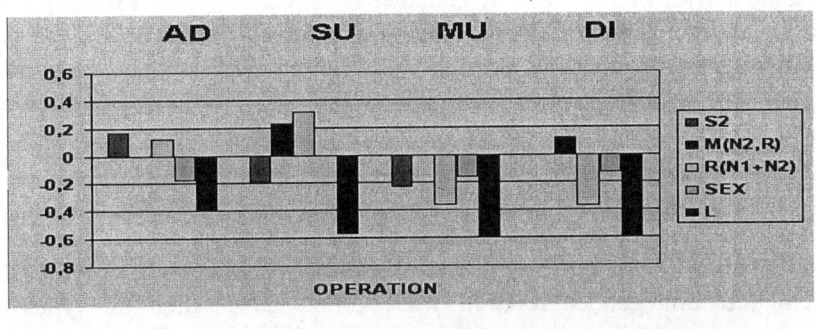

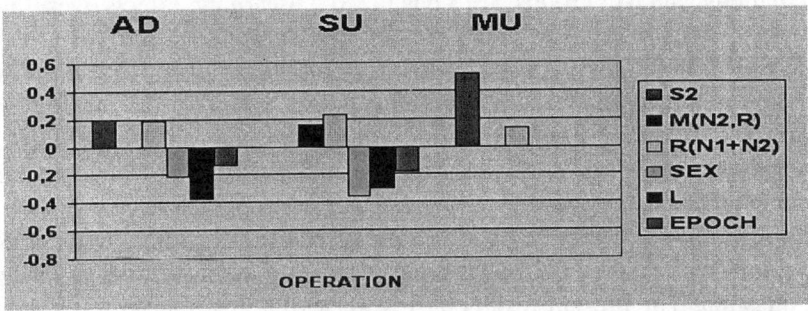

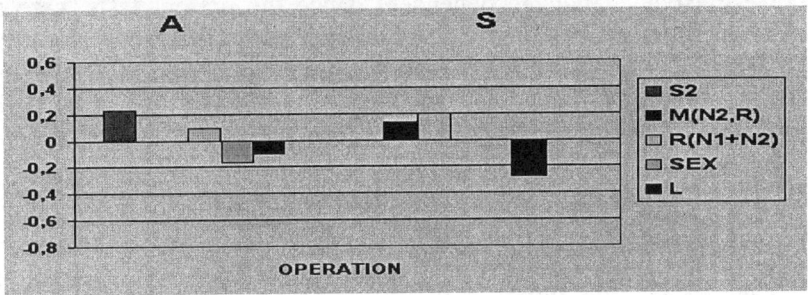

Fig. 8.8. Number size effect: *Calculation time depends on size of the operands. Top: ADU; middle: CHI4 and bottom: CHI2*

The gender differences seem, from the above, to start to appear at the very beginning of academic training in arithmetic (Group CHI2) and to persist into adulthood (Group ADU). It must be remarked that the adult group was composed of graduate students in the exact sciences. This invalidates any strong claim that gender differences in math are the result of a biased education, and seems to point to a phylogenetic explanation.

A common adult commentary at the end of the experiment was that they tried to find a good strategy to quickly solve the tests. This is in clear agreement with the findings of Gallager et al. (2000), and it is supported here by another clear result: dependence of calculation time on the order of question presentation. The strongest correlation observed in the present experiments was the inverse relation between calculation time and the order of the question in the test series, for all type of calculations and for all experimental groups. This clearly shows that volunteers experienced some sort of *learning* while solving the tests. Interesting also is the fact that regression analysis shows that the rate of this learning is greater for males than females. It must be remembered that the correlation coefficient for the calculation time/gender relation was around 20% when all types of calculations and all groups are considered. Learning effects were also demonstrated by Dehaene et al. (1999) and Spelke and Tsvikin (2001), who trained bilingual adults in solving arithmetic calculations in one language and then tested them in both languages.

Some female adult commentaries seem to corroborate the hypothesis of strategy flexibility and its language dependence. Some female adults comment that for them is difficult to use, e.g., the commutative property of multiplication **a * b = b * a,** which may speed up calculation by using the largest multiplicand as the block to be repeatedly summed over to answer the question. The present results clearly show that learning in test solving is not exclusively language dependent, since the present tests were all visually encoded and decoded. Taken together with Dehaene et al. (1999) and Spelke and Tsvikin (2001), these data point to an important property of the neural math circuits, viz., their plasticity, a condition necessary to explain how "number discovery," imposed by the complexity of social transactions, may be stabilized and incorporated into different human cultures. The learning effect observed in the CHI4 children is further strong evidence for such a hypothesis. The calculation time and error rate decreased in this group when the test performances from two different epochs (six months apart) were compared, but this group had not mastered division yet.

But choice and optimization of strategies requires variability of tools to solve the same problem. This property is basic to DIPS and to the model proposed here. Number representation, counting, and calculations are supported by different neural circuits involving neurons located at different brain regions and specialized for different kinds of tasks: cardinal and ordering representation, fuzzy and crisp numbers, addition, subtraction, multiplication, and division.

8.3.2 Size Number Effect

The use of different calculation strategies is confirmed by the regression analysis of the calculation time dependence on number size. As already showed and proposed by others (Fayol, 1996; Fink et al. 2001; Gallistel and Gelman, 1991; Gelman and Meck, 1983; Groen and Parkman, 1972; Jensen et al. 1950; Kaufmann et al. 1949; Mix, 1999; Siegler, 1996), the present results indicate that both children and adult use different strategies to solve each of the arithmetic calculations.

The size number effect was first observed in the group CHI2. Addition time increased as both the minima of the operands (minimum counting simulation) increased (S2 in Fig. 8.6) and the square root of their sum augmented (R(N1 + N2) in Fig. 8.6). Subtraction time increased as both the minima between the subtrahend (counting up simulation) and the result (counting down simulation) increased (M(N2, R) in Fig. 8.6) and the square root of the operands augmented. Similar results in adding and subtracting were also observed in the case of CHI4 children. Furthermore, the multiplication time in this latter group increased as the minima of the multiplicands increased (S2 in Fig. 8.6) and also as the square root of their sum augmented.

The adult addition time increased as both the minima of the operands increased (S2 in Fig. 8.6) and as the square root of their sum augmented (R(N1 + N2) in Fig. 8.6), too. Their subtraction time also increased as both the minimum between the subtrahend and the result (M(N2,R) in Fig. 8.6) and the square root of the operands (as observed in the CHI4 group) augmented. However, their subtraction time decreased as the square root of the sum of the operands increased. The product time decreased as both the maximum multiplicand (S2 in Fig. 8.6) and the square root of the sum of the multiplicands increased. Finally, division time increased as the minima between the divisor (division as inverse multiplication strategy) and the result (repeated subtraction simulation) increased (M(N2,R), as in Fig. 8.6) and the as square root of the sum of the operands decreased.

The results with the CHI2 and CHI4 groups demonstrate that children used full and optimized simulations for solving addition and subtraction problems. Their calculation times were dependent on the square root of the sum of the operands, and also obeyed the optimization strategies of minimum counting for addition, or counting up or down in subtraction (Siegler, 1996). But the size effect dependence never resulted in high correlation coefficients; they remained at around 20%. This means that much of the calculation time variation may be due to the use of other strategies, such as those discussed here as formal calculations and which are supposed to be faster than simulation solutions. The CHI4 children also used different strategies to solve the multiplication tests, since the calculation times depended on both the square root of the sum of the multiplicands and on the size of the smallest of these operands. The 40% correlation in the later case may be an indication that block counting (or counting by multiples) was a predominant strategy over full simulation and formal calculation. The analysis of the videos clearly shows changes of strategies. Sometimes the same children used finger counting, while at other times they finger-pointed figures on the computer screen. In some instances the same children used both hands for finger counting, while on other occasions he/she used one-hand simulation. Most children used repeated movements of finger blocks to simulate multiplication. Similar results were obtained with the twelve students of the second private school, although some correlations did not attain statistical significance at the level $p < .5$, possibly because of the small number of volunteers compared with CHI2 or CHI4. Some of the results with those children in public schools were similar also to those obtained with the CHI4, but again the statistical significance of the correlation was more variable due to the fact that the children in this group were enrolled in many different school semesters.

Adults used different strategies for solving each kind of arithmetic calculation, too. Since the size effect dependence was more complex for adults in comparison with the children, it may be proposed that learning enriches arithmetic knowledge by increasing the number of available strategies for the same calculations. Adults persisted in using the same childhood strategies, since similar calculation-time/number-size relations were disclosed when both groups were compared. Nonetheless, some unexpected inverse dependencies of calculation time with the square root of the operand sum were observed for multiplication and division. Also, calculation was faster if the maximum operand increased. All of these point to the use of other types of strategies to solve these types of calculations. At least one volunteer reported ruler simulation, which is based upon visual operations with the number line (Zorzi et al. 2002).

8.4 The Distributed Mathematical Brain

The MCC and FMs associated with each arithmetic task corroborated the male/female and adult/children differences in performance described in the preceding sections, and disclosed some interesting characteristics of the brain activity associated with the arithmetic calculations.

The adult MCCs were very similar for males and females (Fig. 8.9). High values of $h(d_i)$ for the frontal and central electrodes where obtained bilaterally in the case of addition and subtraction, and mostly over the left hemisphere in the case of multiplication and summation. Different from the MCCs, the factor mappings FM1, FM2, and FM3 greatly differed between sexes. The adult MCCs were very similar for both male and females and for all types of calculation (Fig. 8.9). They show that anterior and bilateral areas were strongly committed to solving any type of arithmetic calculation. The three FMs show, however, that these anterior areas enrolled other neurons (distributed over the entire brain), in different ways when male and females were considered.

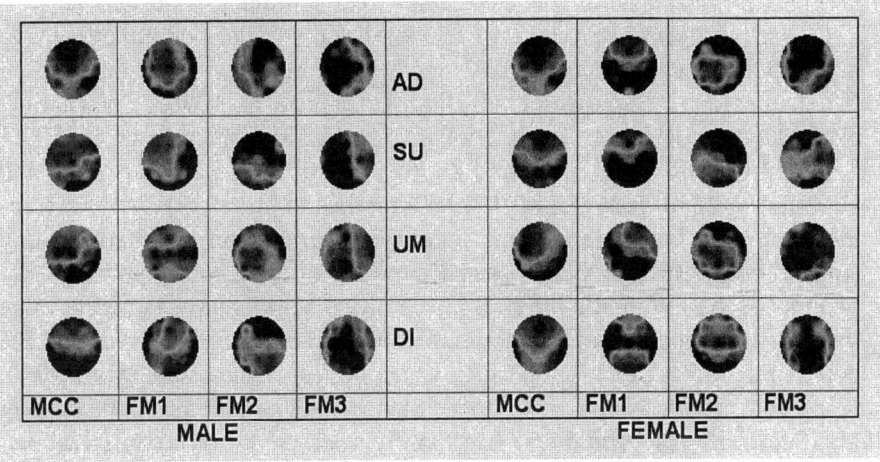

Fig. 8.9. Adult brain mappings. *MCC: $h(d_j)$ Mean mappings; FM1, FM2 and FM3: Factor Mappings generated by Principal Components Analysis. AU: Addition; SU: Subtraction; MU: Multiplication; DI: Division. All mappings are normalized into the closed interval [0,1].*

In the case of addition and subtraction (Fig. 8.9), we propose a set of:
1. left frontal neurons (N_f:): formed by cells recorded at FP1, F3 and FZ;

2. bilateral central-parietal (N_{cp}) cells: formed by neurons recorded at C3, CZ, C4, P3, PZ and P4, and
3. neurons (N_v): distributed more laterally in both hemisphere, which are involved in solving calculations in adults.

On the one hand, the female N_f is more anterior than the male N_f and the latter is more lateral than the former. On the other hand, N_{cp} may be assumed to be more similar for both genders, although sharing an association with N_f that is strong in the case of man and almost absent in the female. Male MF3 displays a strong correlation among areas in the right hemisphere, whereas MF2 points to associations between sites in the left hemisphere and the posterior cortex. Female MF3 display patterns similar to male MF3 and includes also some of the areas appearing in man's MF2.

Also, the N_v neurons are more correlated in males than in females. Finally, it may be assumed that N_f and N_{cp} are sets of neurons involved with task solving, whereas N_v is in charge of visual computations tasks.

Frontal and central-parietal, or temporal and parietal components, have being described in many fMRI studies on addition and subtraction (Cochon et al. 1999; Cowell et al. 2000; Dehaene et al. 1999; Jahanshahi et al. 2000; Menon et al. 2000; Rickard et al. 2000; Stahian et al. 1999; Stanescu-Cosson et al. 2000; and Zago et al. 2001) and ERPs (Iguchi and Hashimoto, 2000; Kong et al. 1999; Skrandies et al., 1999). All these authors reported widespread areas involved in arithmetic calculations, and stressed both left frontal and parietal areas as common and important components of the arithmetic brain. Some of these authors have suggested that the frontal component of such circuitry (our N_f) is much involved with the complexity of calculation, besides other duties (Jahanshahi et al. 2000; Kong et al. 1999; Menon et al. 2000; Stanescu-Cosson et al. 2000; Zago et al. 2001). The N_{cp} component is generally assumed to have a bilateral distribution and a specific role for arithmetic computations, and its activity is described as mostly dependent on the type of calculation and on number size (Cochon et al. 1999; Cowell et al. 2000; Dehaene et al. 1999; Menon et al. 2000; Rickard et al. 2000; Stanescu-Cosson et al. 2000). Some authors have also referred to other visual and verbal components associated with arithmetic calculations that involve other neural circuits (e.g., Cowel et al. 2000; Dehaene et al. 1999; Zago et al. 2000). Among all the papers listed above, only Skandries et al., (1999) have reported that females consistently have larger global field power in EEG than males, and they also displayed different scalp field topography for various components. These authors also stressed that early visual processing ERP components were gender sensitive. The gender differences in calculation time for both addition and subtraction (Rocha et al., 2003c, d) may be explained by the use

8.4 The Distributed Mathematical Brain 173

of different strategies by men and women, reflected in a better coordination between N_f and N_{cp} (as shown by FM1) in the case of men than the association disclosed by MF1 and MF2 in the case of women. Also, these differences appear to be supported by a different N_v enrollment in both male and female task solving.

Now, let us propose the N_{cp} as the set of neurons in charge of implementing the set of accumulators A and the KFN and CBN circuits (see Chap. 7), whereas N_f is the set of neurons implementing both the ordinal numbering control of the counting pathway (Rocha, Rocha and Massad, 2003c, d) and the gate control of the KFN and CBN circuits exercised by the C neurons, one of the components of N_f. If this is the case, solutions to addition and subtraction problems may be arrived at by using the KFN and CBN circuits, and by means of both simulation and formal calculation, as proposed by Rocha et al., (2003c, d). In the case of formal calculation, the visually displayed operands are recognized by visual neurons (R neurons; see Fig. 7.6) that project directly to the corresponding I neurons. In this way, e.g., 4 + 3 or 7 – 3 involves the decoding of the numerals 4, 7, and 3 by some specific neurons in the visual associative cortex and their semantic evaluation by means of the I neurons. The type of calculation (+ or -) is visually recognized and semantically decoded by other neurons of the N_f set, since the C neurons are in charge of the different calculational simulations proposed by different authors in the literature (Butterworth, 1999; Dehaene, 1991, Dehaene, 1997; McCloskey et al., 1991; Siegler, 1996) . In the case of simulated operations, numbers representing quantities may first be decoded as set of elements that are then used to mentally simulate the process of counting (up or down) of the sets representing the operands. Hybrid calculations may be processed by loading one of the numbers directly into a $\in$ A, using the corresponding i $\in$ I and decoding the other number as a set of elements, whose elements are sequentially accumulated into the same a $\in$ A through the adequate gate g $\in$ G. If the operands are first ordered, then this hybrid calculation may be optimized by loading directly the highest operand and decoding the other as a set of elements. Thus, both the KFN and CBN may allow both males and females to use distinct adding/subtracting strategies (e.g., Fayol, 1996; Gelmann and Gallistel, 1991; and Siegler, 1996), which in turn shapes their N_f and N_{cp} and results in their distinct MFs. It is easily verified that simulated calculations may render addition time dependent on the size of operands, whereas formal calculation makes it constant. Since different sets of agents may enroll to solve the same task proposed to a DIPS, then it may be proposed here that men/women may use different strategies of recruiting the KFN and CBN for their calculations, which could explain the gender differences observed for the calculation response time (Rocha et al., 2003c,d) and for the MFs.

Data on multiplication may be easily understood according to the proposed model, too. Both male and female FM2 show that N_{cp} neurons are engaged in this task, being left (male FM1) or right (female FM1) controlled by the N_f circuit. Many fMRI studies have also disclosed both frontal and parietal components associated with multiplication (Cochon et al. 1999; Rickard et al. 2000; Skrandies et al., 1999; Zago et al. 2001). FM3 (associated with N_v) shows a strong correlation among right hemisphere areas in the case of men, whereas it is almost absent in women. Again, the neuronal enrollment is different between sexes. Counting by block (or multiple counting) may explain product solving (Rocha et al., 2003c,d). Either formal block counting is left-controlled in the case of men, or block counting is visually (right) simulated in women. This could explain males being faster than females, although the difference is the smallest observed for all calculations. This small difference may be accounted for by the heavy dependence of multiplication tasks on MF2 being equally associated with N_{cp} neurons for both genders. It must be remembered that also Skrandies et al. (1999) pointed out that scalp field distributions were affected by gender, indicating the activation of different neuronal assemblies during visual information processing of males and females.

Division is the operation that most differed concerning the factor mappings. In the case of men, MF1 showed a strong correlation for FP2, F8, FZ, and CZ, whereas MF2 displayed an association between F3, C3, P3, PZ, P4, O2, and OZ. It also resulted in a bilateral pattern of MF3. To the best of our knowledge, only Skrandies (1999) have also included division among the calculations proposed to their volunteers, but these authors did not describe their results as taking into consideration the distinct types of calculation. Perhaps adults make more use (as reflected by MF3) of verbally encoded rules of thumb (e.g., all even number are divided by even numbers, etc.) to orient calculation. On the one hand, perhaps males used this type of information to better control the N_{cp} neurons distributed bilaterally (see male division MF2 in Fig. 8.9), or to perform other visual formal operations (male division MF1 in Fig. 8.9) such as line number sectioning (Zorzi et al., 2002) as reported by at least one volunteer. On the other hand, females may have used the same type of information to orient different counting up/down strategies to solve division, since their cerebral patterns for this operation have some similarity to their subtraction MFs if one considers that most of the subtraction MF2 is present in MF1 division, and that there is some resemblance between subtraction MF3 and division MF2. This may be an indication that females made more use of the multiple counting up/down strategies discussed by Rocha et al. (2003c) and may explain why division calculation time differences between men and women was the second most significant.

8.5 Building the Distributed Mathematical Brain

The CHI4 and ADU brain mappings for addition and subtraction were more similar than those for multiplication and division. But children (see Fig. 8.10) were less trained in these later calculations than on addition and summation. Their addition and subtraction calculation times are statistically smaller than the corresponding multiplication and division times, as well (Rocha et al., 2003c). It is also interesting to note that children used a lot of finger pointing and marking while solving each arithmetic question, whereas adults used mental simulations instead of these overt manipulations for the same purpose. This could be the main source of the adult/children brain mapping differences and explains the fact that children are slower and err more than adults in doing any kind of arithmetic calculation.

CHI4 and ADU male addition mappings share more similarities than those for females. The boys' MCC includes the adult cerebral areas plus a moderate activation over P4, T5, and OZ; their MF1 disclose a correlation pattern very similar to the adult one, and their MF2 may be assumed to be a combination of the adult MF2 and MF3. By contrast, female mappings differed between adults and children, if MF3 is not considered. High $h(d_j)$ values were obtained for the CHI4 girls over C3, FP1, F4, FZ, C4, PZ, and P4 and these areas seemed to be well correlated, as disclosed by MF1. Also, F3, C3, and O1 are the only high correlated areas in children's MF2. The similarities between children and adults are even greater in the case of subtraction, although the gender differences appear to be greater between boys and girls than between men and women. These differences between boys and girls, and the similarities between children and adults, may be understood if children are still developing N_f and N_{cp} sets, and boys are more advanced than girls in building them. This could explain boys being faster than girls in arithmetic calculation.

There are more similarities between girls' multiplication and division mappings than between the corresponding women's or boys' mappings. Also, these arithmetic mapping differences were clear between men and boys. Taken together, the MFs disclose patterns of greater coordination among enrolled areas in the case of boys as compared to girls, and in the case of adults as compared to children. Also, the MCCs seem to provide evidence for neuronal recruitment in solving multiplication and division that is greater for girls than for boys and adults. All of these may be understood as multiplication and division in children's N_f and N_{cp} buildup is more primitive than their corresponding modeling of addition and multiplication, a fact that parallels the differences in the amount of their training on these different arithmetic operations.

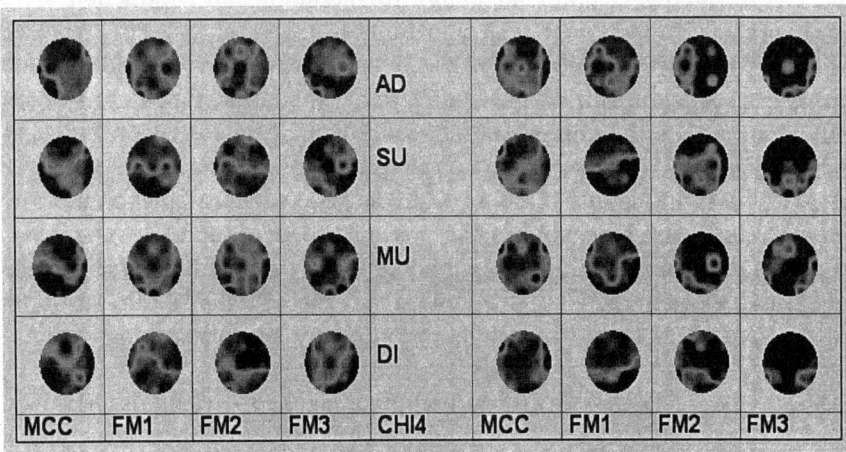

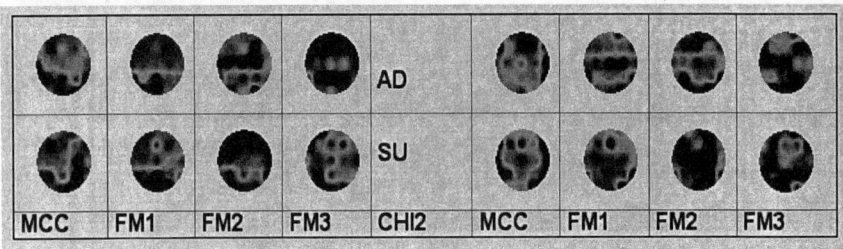

Fig. 8.10. Children brain mapping: *MCC: h(d$_j$) Mean mapping; FM1, FM2 and FM3: Factor Mappings generated by Principal Components Analysis. AU: Addition; SU: Subtraction; MU: Multiplication; DI: Division. CHI4: Children enrolled in 4th and 5th semesters of elementary school; CHI2: Children enrolled in 2nd and 3rd semesters of elementary school. All mappings are normalized into the closed interval [0,1], and color encoded according to the rule: 1 is red and 0 is black.*

Finally, although there are some similarities among the CHI2 and CHI4 MCCs, the differences between their MFs are well defined. But the CHI2 calculation times were also very different from those obtained for the CHI4 group. Again MF patterns seem to correlate with the amount of training experience by and the degree of proficiency attained by children.

8.6 Conclusion

As a general conclusion, it could be said that the main difference between adult and child brain activity patterns, as disclosed by MCs and MFs, is a larger neuronal enrollment among children in comparison to adults, due, perhaps, to a more generalized use by children of simulated, rather than formal counting and calculations. Another difference consists in better adult N_f and N_{cp} circuit development. The search in the literature for papers dealing with arithmetic, learning, and brain activity selected only the work of Menon et al. (2000), an fMRI study about optimization of arithmetic processing in perfect performers. These authors showed that activation of the left angular gyrus was training dependent.

Also, the gender differences reported here, for both adults and children, are very novel data. Although it is a well established fact that boys outperformed girls in the SAT-Math exam, and it has been proposed that test-solving speed and strategy flexibility may be explanations for such a finding (Royer and Tronsky, 1999, Gallager et all, 2000), the present work seems to be one of the first attempts to provide an ample analysis of the possible brain activity differences between genders concerning arithmetic cognition. To our knowledge, only Skandries, Reik and Kunze (1999) have reported that females consistently have larger global field power in EEG than males, and they also displayed different scalp field topography of various components. These authors also stressed that early visual processing ERP components were gender sensitive. The present results indicate that marked differences exist between sexes in brain processes supporting arithmetic calculation, and that these differences are present from the very beginning of academic training. Also, differences were found for male and female adults that experienced a more similar and advanced mathematical training due to the fact that they are successful postgraduate students in the field of technology. All these facts seem to point toward an important phylogenetic component of these gender differences that may in part be augmented by a gender influenced culture, which also has phylogenetic roots.

9 Arithmetic Learning Capability in Congenitally Injured Brains

The arithmetic learning capability of brain-damaged children with low and normal IQ is studied and discussed, taking into account the model proposed. The results clearly demonstrated that:

1. widely distributed bilateral parietal lesions reduces the children arithmetic learning capability;
2. left frontal lesions may dissociate the capability of handling and operating quantities from that of reporting these results;
3. intermittent delta rhythmic activity is associated to dyscalculia in normal IQ children, and
4. brain plasticity allows children to overcome most of their arithmetic learning problems.

Also, the results corroborate the conclusions about the organization of the arithmetic neural circuits discussed in Chap. 8.

9.1 Dyscalculia

The knowledge of numbers has a phylogenetic root (Rocha and Massad, 2002; 2003; Rocha et al., 2003c; Wynn, 1988, 2000) and may be disturbed in a variety of forms, due to different causes. Sometimes children show particular problems with arithmetic. If their intelligence is considered to be normal this is usually called *dyscalculia,* otherwise it is named *numeracy deficit* (Ansari and Karmiloff-Smith, 2002). Gerstmann's syndrome is often described in adults and is caused by acquired lesions, usually vascular lesions or tumors involving the angular gyrus of the dominant parietal lobe; however, it has also been described in children with learning disabilities by the name of *developmental Gerstmann's syndrome* (Mayer et al. 1999; Suresh and Sebastian, 2000).

In a recent review (Suresh and Sebastian, 2000), clinical epilepsy, abnormal EEG findings, and nonspecific MRI changes were associated with developmental Gerstmann's syndrome. Possible contributing factors to dyscalculia may be as diverse as genetic predisposition, neurological abnormalities, and environmental deprivation (Shalev and Gros-Tsur, 2001). The cause of numeracy deficit is not well discussed in the literature, specially in those cases where the intelligence deficit does not have a genetic source but is associated with gross or slight brain damage (Ansari and Karmiloff-Smith, 2002; Rocha et al. 2003d).

Although cognitive neuroscience has greatly advanced our knowledge of numerical cognition, it has neglected the development of strong formal models of brain computations supporting human and non-human arithmetic capability, as well as the study of number in cognitively impaired children (reported e.g. in Ansari and Karmiloff-Smith, 2002; Isaacs et al. 2001; Rocha and Massad, 2002; 2003; Rocha et al.,, 2003c; Rocha et al. 2003d).

Here, data from Rocha et al. (2003d) are summarized regarding arithmetic skill development in a group of impaired children. This data is then correlated with the children's possible brain damage, and further used to support the propositions introduced in Chapters 7 and 8, concerning the composition of cerebral arithmetic circuits.

9.2 Damaged Brains

The boy **WS** (fig. 9.1, 9.2) was born in 1987 from a mother experiencing high blood pressure during her pregnancy. He experienced delay in language development, uttering his first words at the age of one year and eight months, and beginning to produce **SOV** (Subject-Verb-Object) at the age of seven. His IQ is around 60. His MRI revealed a huge lesion – leukomalacy – of the left parietal lobe, probably due to a pronounced increase in his mother's blood pressure that resulted in a week of hospitalization. He began elementary school at the age of ten, and exhibited slow development in his reading, writing, and arithmetic skills.

By the year 2001 he was able to sum and subtract numbers of up to three digits, to multiply numbers of one digit, and to solve simple problems involving summation, subtraction and multiplication, although he used quantity manipulation in many instances. His performance in mathematics was, however, better than in reading and writing. One year later, he improved his addition and subtraction skills, but continued to use manipulation to deal with one-digit multiplication. In 2003 he was attending the last year of elementary school. His 1991 MCCs showed a reduced enrollment of neurons located under FZ, CZ, C3 and C4 for both summation and subtraction. His factor mappings (FM1 and FM1) disclosed two patterns of

$h(n_i)$ covariation, one among areas in the left hemisphere, and the other involving areas on the right side of brain. Despite the huge left parietal lesion, subject 1 seems able to enroll neurons in the left hemisphere in his calculations.

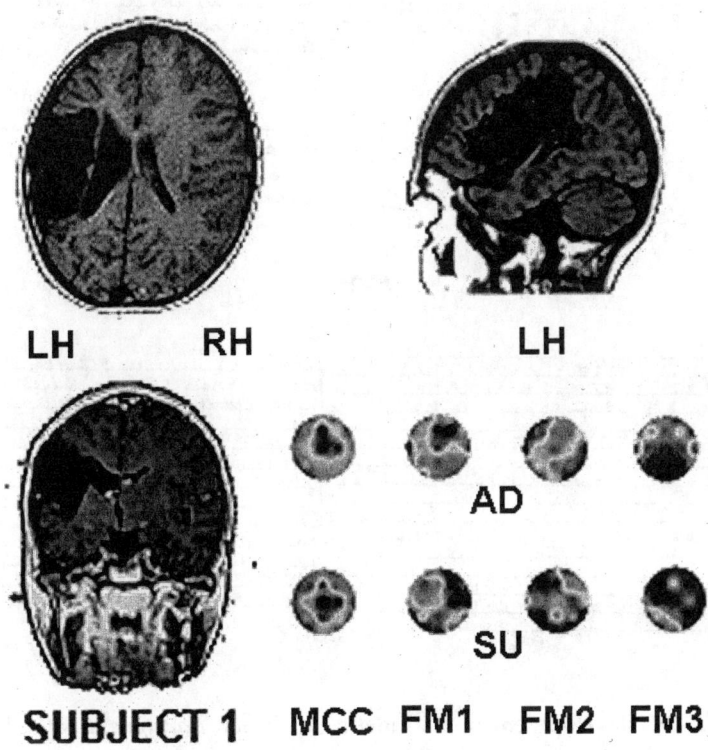

Fig. 9.1. WS, a case of left parietal leukomalacy. *MCC: $h(d_j)$ Mean mappings; FM1, FM2 and FM3: Factor Mappings generated by Principal Components Analysis; AD: Addition; SU: Subtraction. For the MRI: RH: right hemisphere; LH: left hemisphere.*

The girl **KTS** (Fig. 9.3, 9.4) was born in 1986, also from a mother with high blood pressure, and also hospitalized during one month, in the last trimester of her pregnancy. She experienced a delay in motor development, standing up by the first year of age and beginning to walk by the age of one year and nine months.

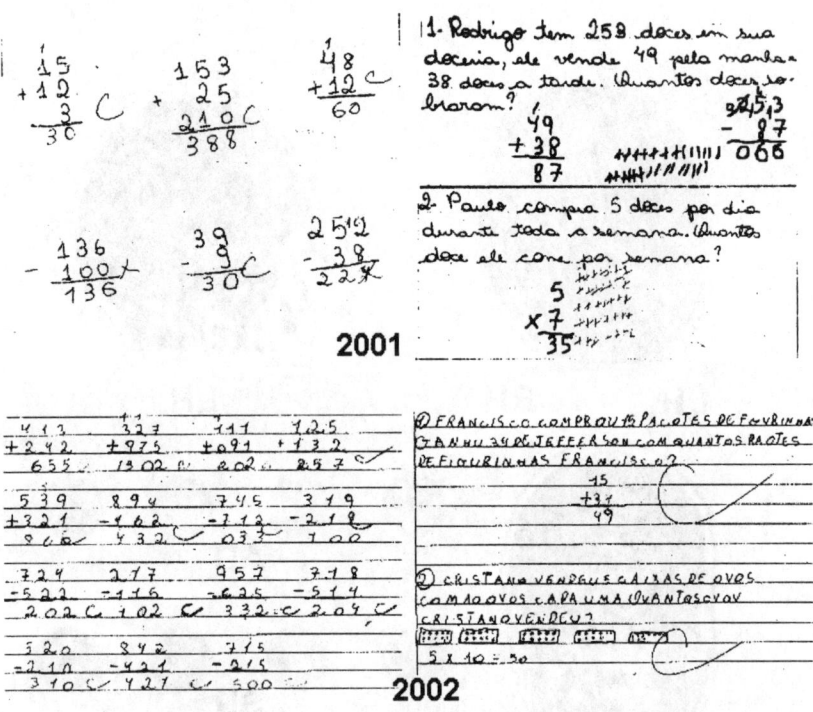

Fig. 9.2. WS development in arithmetic capability. *Portuguese is the language employed in copies of the original classroom exercises.*

The **KTS** MRI revealed a huge lesion of the right parietal lobe, and her present IQ is 65. She began elementary school at the age of nine; her reading, writing and arithmetic were considered adequate, although her language performance was above that for mathematics. Her arithmetic capability by the year 2001 was well advanced, and included summation and subtraction for numbers up to four digits, multiplication of numbers up to two digits, and division by numbers of one digit. She is now attending a special program for young adults in high school education. Her 1991 MCCs also showed reduced enrollment of neurons located under FZ, CZ, C3, C4, PZ and P4 for summation, subtraction and multiplication.

Her factor mappings revealed three patterns of $h(n_i)$ covariation, one among posterior areas in both hemispheres, another among areas in the left hemisphere, and the third one involving areas in the right side of brain. Despite the huge right parietal lesion, she seems to be able to enroll neurons in both the left and right hemispheres in her calculations.

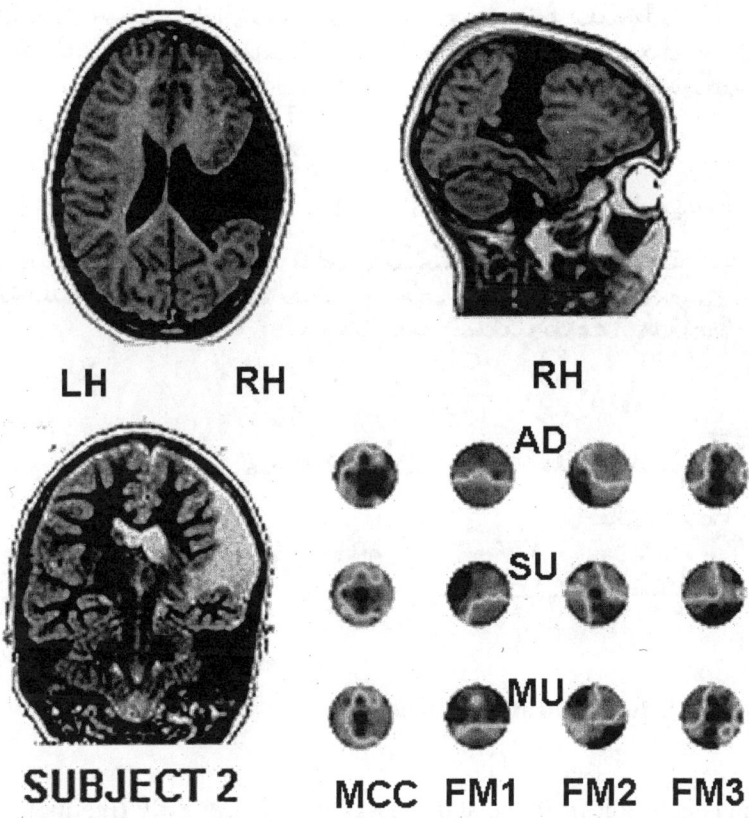

Fig. 9.3. KTS, a case of right parietal leukomalacy.

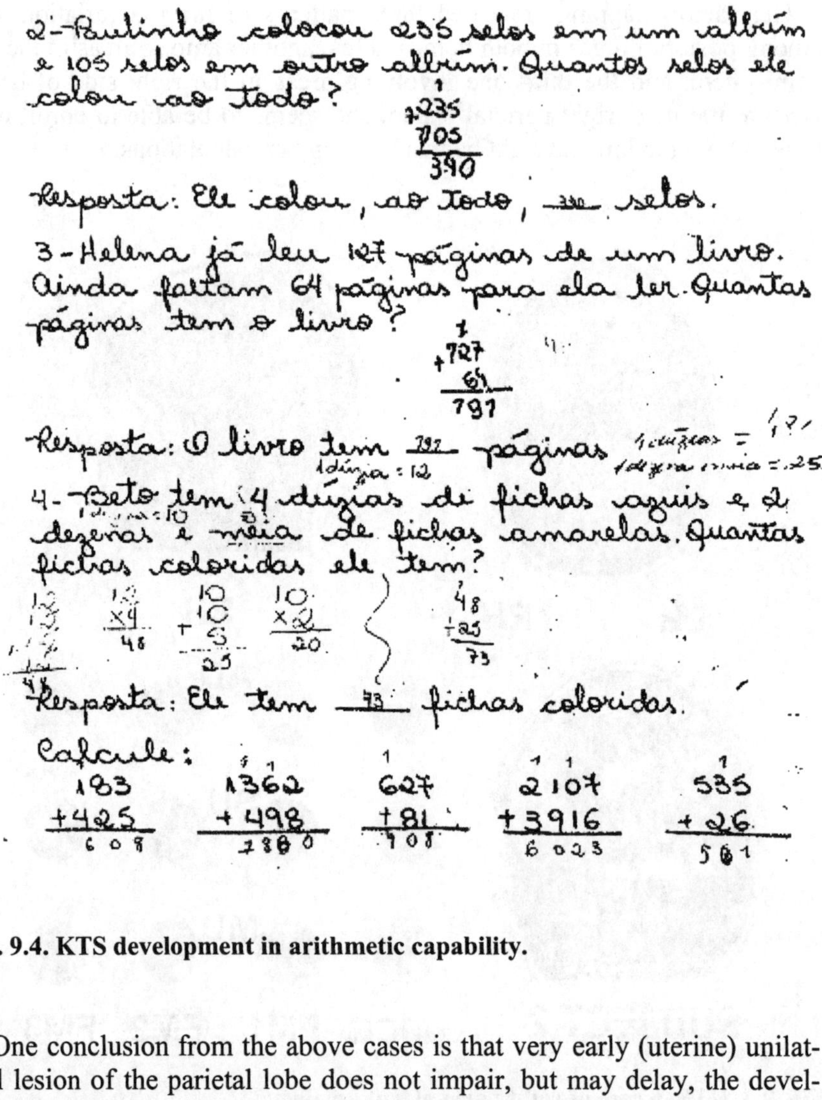

9 Arithmetic Learning Capability in Congenitally Injured Brains

[handwritten student work in Portuguese showing arithmetic problems and solutions]

Fig. 9.4. KTS development in arithmetic capability.

One conclusion from the above cases is that very early (uterine) unilateral lesion of the parietal lobe does not impair, but may delay, the development of arithmetic skills. Thus, it seems that the N_{cp} component of the arithmetic circuits modeled in chapter 7, and revealed by the brain mapping in chapter 8, is fault tolerant, perhaps because of a redundancy of number representation in both hemispheres. The N_{cp} is hypothesized to correspond to the set of accumulators and quantifiers located bilaterally at the parietal lobe, whose activity could be monitored by means of the parietal (P3, PZ, and P4) and central (C3, Cz, and C4) electrodes. The lesion of such a component in one hemisphere may be compensated by increasing the number of neurons with the same specialization in the other hemisphere. Cases 3 and 4 seem to confirm this hypothesis.

Subject 3 (Fig. 9.5) is a boy, with an IQ of 54, born in 1990, premature from a mother with high blood pressure. He experienced motor development delay, standing up by 1 yr 2 mo, and walking by 2 yr, 3 mo. His MRI revealed a huge bilateral lesion of the parietal occipital lobes. He is now attending kindergarten, and is beginning to recognize some letters; he continues to experience difficulties in recognizing quantities above five.

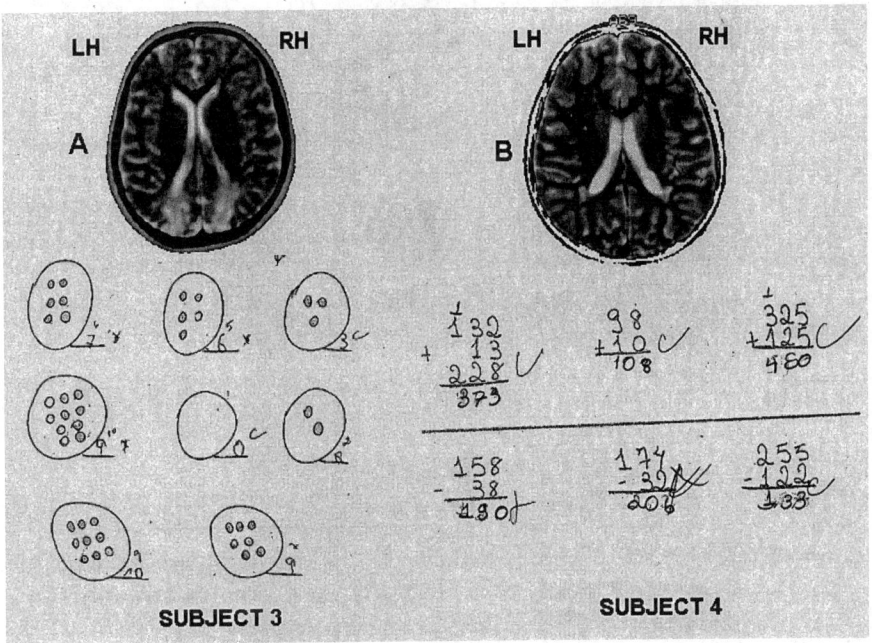

Fig. 9.5. Two cases of bilateral leukomalacy with different arithmetic capability development

On the other hand, subject 4 (Fig. 9.5) is a girl, IQ 61, born '89, with no history of disturbances during the pregnancy. She experienced language development delay; her first phrases were spoken by the age of 3 yrs. Her MRI revealed a small bilateral lesion of the parietal-occipital lobes. She began elementary school in 1999, and is considered to have a regular development in her reading and writing capabilities, but shows a better acquisition of arithmetic skills. She is now able to perform additions and subtractions of numbers of up three digits, but is experiencing important difficulties in learning multiplication. Her brain mappings indicated the enrollment of neurons at the parietal electrodes in both hemisphere in the course of solving addition and subtraction problems. Thus, a huge bilateral posterior lesion resulted in a greater arithmetic capability than unilateral or small bilateral damage.

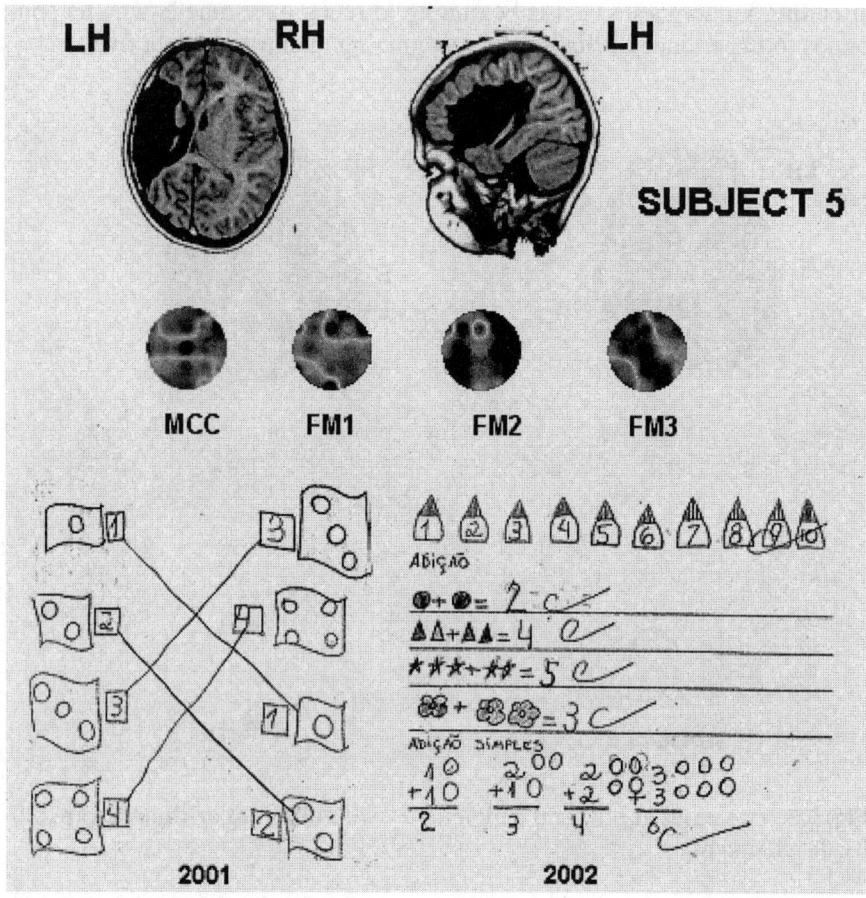

Fig. 9.6. A case of left parietal-frontal leukomalacy and arithmetic capability development

The histories of frontal damaged children are different. The boy **RC** (Fig. 9.6) was born in '95 and his present IQ is 57. There is a history of attempted abortion here, resulting in a huge frontal-parietal lesion. Because of this, he experienced a motor development delay, standing up the age of

1 yr, and taking his first steps at 1 yr, 6 mo; he is now attending the second year of elementary school. He began to cope with quantities above five in 2001; at the same time, he started logographic reading of a small set of words referring to animals. He continued to improve his performance during 2002, and began to add and subtract small numbers. His 91 MCC, associated with quantity and number recognition, showed low enrollment of neurons in F7, T3, C3, CZ, which could be explained by the location of his lesion. Despite this, he was able to recruit both left frontal (see FM1) and parietal (see FM2) neurons to solve quantification tests.

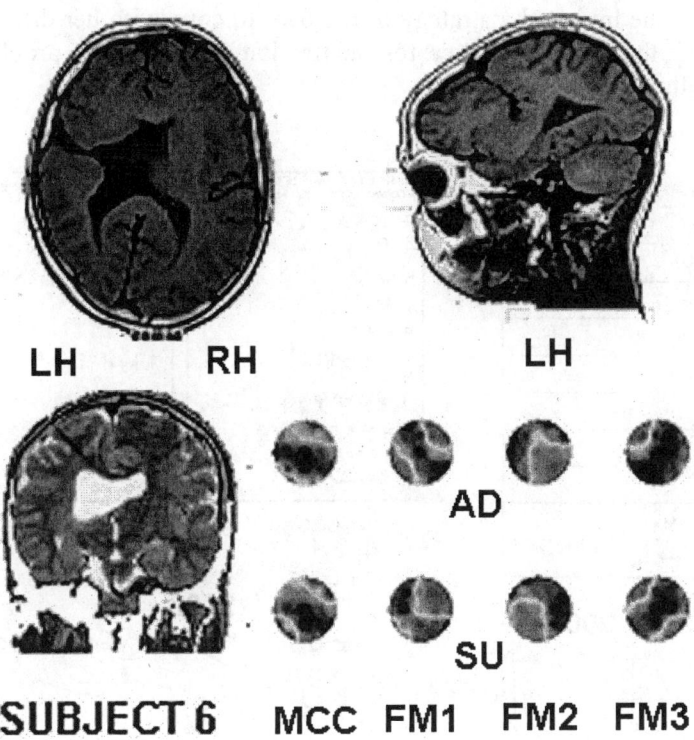

Fig. 9.7. A case of left frontal schizencephaly. MCC: $h(d_j)$ Mean mappings; FM1, FM2 and FM3: Factor Mappings generated by Principal Components Analysis; AD: Addition; SU: Subtraction; for the MRI: RH: right hemisphere; LH: left hemisphere.

The girl of Fig. 9.7 (Subject 6) has an IQ of 58; she was born in '88 from a hypertense mother who experienced one week of hospitalization during the first trimester of pregnancy. This subject experienced motor development delay, standing up by the age of 1 yr 3 mo, and taking her first steps by the age of 1 yr 7 mo. She also experienced an important language development delay. Her first words were spoken at the age of five. Her MRI revealed a huge frontal lesion named schyzencephaly, a consequence of a fetal isquemia during her mother's one week of hospitalization. She began to attend elementary school in 2000, and exhibited good learning in reading and writing since then. She is now able to read small texts and write small notes. Her development of arithmetic skills was almost nill until 2002, when she invented a strategy of her own to cope with her difficulty in selecting the correct answer for subtraction and addition problems, which results were above seven.

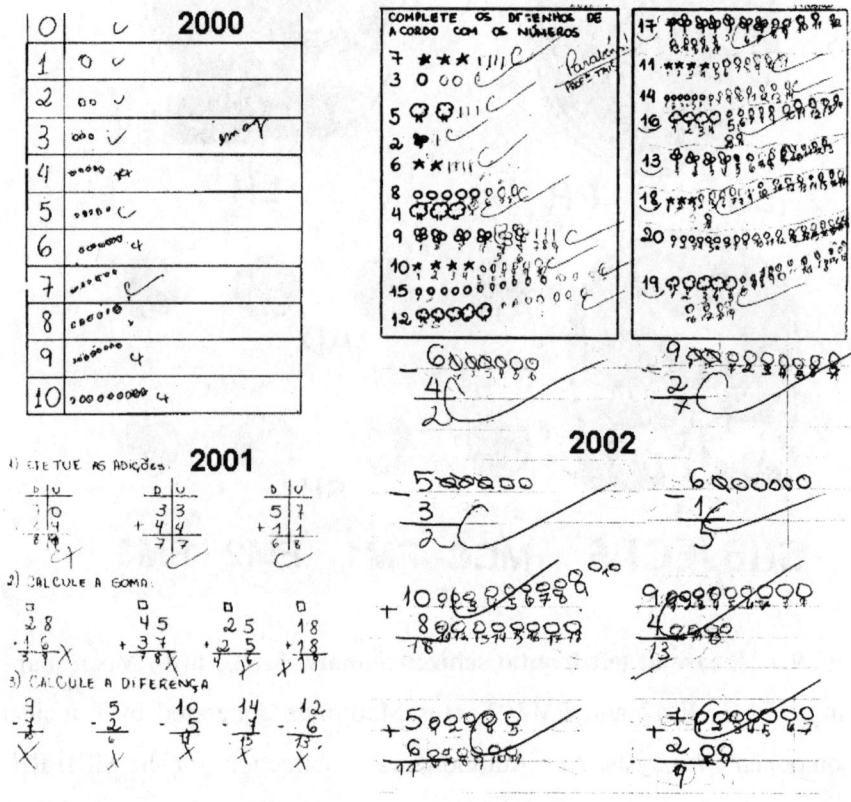

Fig. 9.8. Arithmetic capability development of Subject 6

Whenever she has to report a result of any arithmetic problem that is bigger than seven, she represents the quantities and writes the corresponding series of numbers in order to find the answer. It seems that her problem lies in encoding the results of the quantification, and not in operating on the quantities. Her 2000 MCC seems to point to a difficulty in enrolling frontal neurons for both addition and subtraction, although her FM3s exhibit patterns of some association of these frontal neurons with posterior cells. However, FM1 and FM2 showed an important involvement of right frontal neurons in both addition and subtraction.

Damage to the frontal component N_f seems to delay, but not to seriously compromise, the capability of the brain to quantify the cardinality of sets by using the N_{cp} circuits. It seems that frontal lesions result in a complex deficit of handling ordinal numbers and encoding quantities into numerals. These two frontal-damage children use a lot of simulation and manipulation to solve their mathematical problems, what may be an indication that they have to turn to frequent visual simulations controlled by the intact right brain, instead of using a better left-frontal control of the N_{cp} circuits. This reduced left-frontal counting control may be the reason for using an explicit process of numbering the elements of the sets added or subtracted, in order to find the numerals representing the result of the calculation performed by using N_{cp} simulation.

This hypothesis seems to be confirmed by data from subjects with small frontal lesions (as in Fig. 9.9). This girl was born in '87, with no history of problems during the pregnancy, and no important neuropsicomotor delay. Her MRI revealed a frontal-parietal gliosis and thickening of the posterior corpus callosum. She began elementary school in 2000, and experienced good learning in reading and writing. Her arithmetic learning is progressing, and at the moment she has mastered addition and subtraction of two-digit numbers and multiplication of one-digit numbers. However, she experiences difficulties with the ordinal numbers above twenty. She has a clear dissociation between quantification evaluation/manipulation and number sequence ordering. Contrary to subject 6, she does not have any problem in labeling the results of addition and subtraction. Her 2000 MCC displayed clear patterns in use of frontal and parietal neurons, and of a correlation of activity between them.

It is proposed in the literature that parietal number representation involves both hemispheres (Cochon et al. 1999; Cowell et al. 2000; Dehaene et al. 1999; Fink et al. 2001; Gobel et al.; 2001; Menon et al. 2000; Rickard et al. 2000; Stanescu-Cosson et al. 2000; Zorzi et al., 2002). This could explain why subject 3 is still very compromised in his arithmetic capabilities, whereas subjects 1, 2 and 4 have achieved advanced calculation skills. Both frontal and parietal lesions delayed the learning of arithmetic operations, but frontal damage continues to compromise the handling of ordinal numbers in subjects 6 and 7.

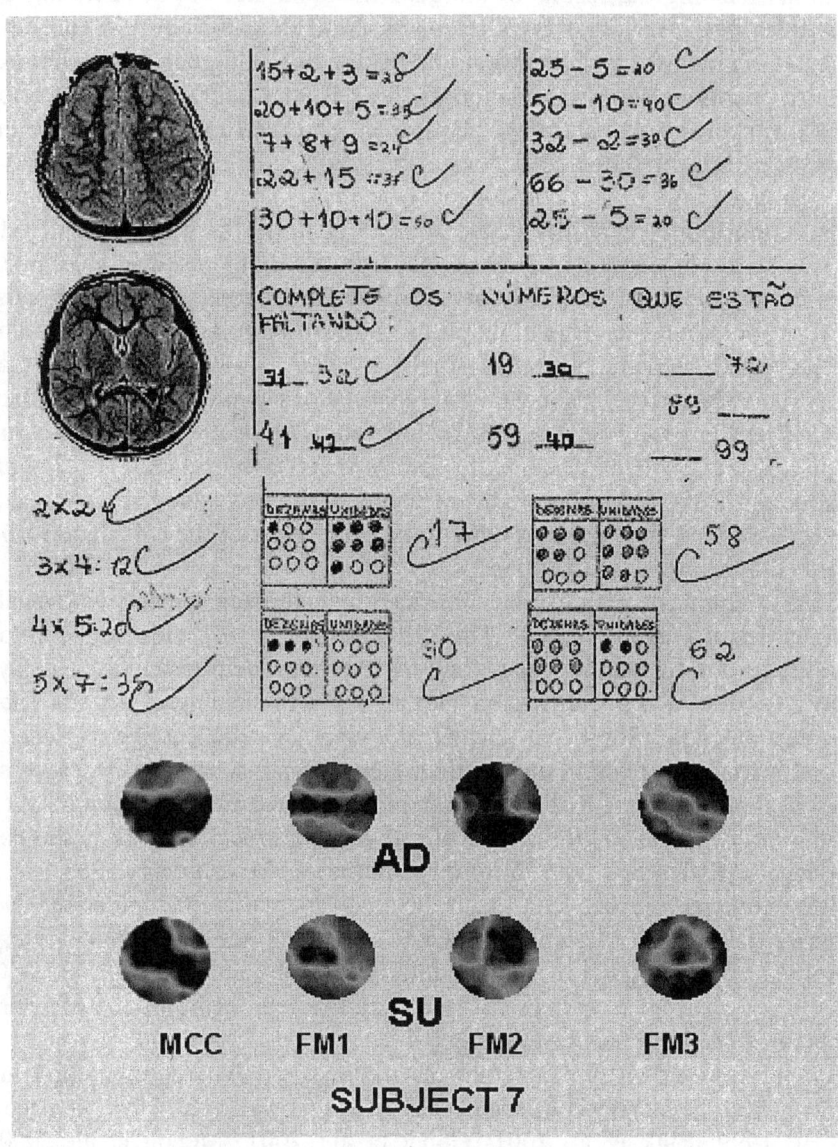

Fig. 9.9. A case of frontal-parietal Gliosis and arithmetic capability developmen

This is consistent with the description of ordering-sensitive neurons in the monkey motor cortex (Carpenter et al. 1999). Finally, it has been suggested that the frontal component of such circuitry is much involved with the complexity of calculation, in addition to other duties (Jahanshahi et al. 2000; Kong et al. 1999; Menon et al. 2000; Stanescu-Cosson et al. 2000; Zago et al. 2001).

9.3 Neural Plasticity

It is clear from the above data that arithmetic capability is preserved in those children suffering from huge cerebral lesions acquired during their fetal life. The unilateral destruction of either the frontal or parietal lobes (subjects 1, 2, 4, 5 and 6) resulted in an important development delay in both language and arithmetic skills, but did not preclude these individuals from attaining sophisticated capabilities in reading, writing and calculating. The language development of these children (subjects 1, 2, and 6) were described and discussed in a previous paper (Foz et al. 2001), and neural plasticity was assumed in the explanation of the reorganization of the language neural circuits, allowing these children to achieve high levels of performance in speaking, reading and writing. Here, the same hypothesis may be used to explain the reorganization of the arithmetic brain in the fetal damaged brains. Suresh and Sebastian (2000) also reported that intensive training improved the arithmetic capability of 60% of their subjects exhibiting developmental Gertzmann Syndrome. It is interesting to consider that the bilateral parietal lesion of subject 3 severely impaired his learning with regard to both numbers and letters. However, the bilateral parietal lesion necessarily severely reduced his neural plasticity, too.

Neural plasticity requires both function reassignment of the neuron and establishment of adequate neuronal connections (Chugani, 1999; Thompson et al. 2000; Villablanca and Hovda, 2000). The latter may be achieved in a densely connected, distributed processing system like the brain, and the former may be understood if neuronal function is dependent on genetic instructions shared by all cells. Rocha et al. (2003) proposed that the specialization of the different types of neurons (accumulators, classifiers, controllers, etc.) involved in the arithmetic circuits may be explained by the control over genes defining the types and quantity of ionic gates, as discussed in Chap. 7. For example, continuous or periodic accumulating functions may be specified by favoring tonic or phasic spike encoding by the piramidal neurons (Rocha, 1997). The expression of the genes required to specialize the arithmetic neurons may be assumed to be phylogentically specified in the case of fuzzy numbers (KFN circuit in Rocha and Massad, 2003a) and culturally guided in the case of crisp numbers (CBN circuit in Rocha et al., 2003a). This assumption helps in understanding both brain plasticity and the fact that man has independently created (or discovered) the crisp numbers at least four times in his evolution — by the Summerians, Epyptians, Greeks, and Mayans (e.g., Ifrah, 1985). Whenever the environment demanded it, genes responded, creating new types of arithmetic neurons. In a similar way, the complexity of commercial or scientific (e.g., astronomical) transactions pressed for the invention of crisp numbers, although adequate teaching and challenging may guide brain plasticity. All subjects 1-6 attend a special education school where teaching is oriented by the recent knowledge provided by neuroscience about language and arithmetic cerebral circuits (www.enscer.com.br). As a matter of fact, their arithmetic teaching was both guided and inspired by most of the research described here and in Rocha et al. (2003c,d).

9.4 Developmental Dyscalculia

Dyscalculia in school-aged children has received less attention in the literature than language and literacy deficits (Ansari and Karmilof-Smith, 2002; Suresh and Sebastian, 2000). Also, number knowledge disruption is much more frequently studied in adults than in children (e.g., Butterworth, 1999; Dehaene, 1997; Fink et al. 2001). Subjects 8 (Fig. 9.10a,b) and 9 (Fig. 9.11) are examples of children having difficulties in learning arithmetic, reading, and writing.

On the one hand, we have subject 8, a boy with a normal IQ, born in '97, from a normal pregnancy, and with a normal childhood history. He began elementary school in 2002, when he began to exhibit important difficulties in reading words, associating numbers with quantities, and performing addition, although he was clearly capable of recognizing quantities when asked to compare sets with equal or different cardinalities. He ended the school year being unable to show any real progress in reading or arithmetic. His 2003 MRI disclosed a pattern of delayed myelinization. His EEG was recorded twice in 2000 and revealed in both instances the presence of Intermitent Rhythmic Delta Activity. His **FMs** disclosed only two patterns of correlation of cerebral activity for quantity and number association (**NU** in Fig. 9.10a), and for addition of small numbers (**AD** in Fig. 9.10a), instead of the usual three Factor Mappings associated with our tests, and also exhibited by in him in the case of set cardinality comparison (**EQ** in fig. 9.10a).

On the other hand, there is subject 9, also of normal IQ, a boy born in 1994 and adopted by the age of four. He had a history of child abuse prior to this date. He began (and continues) to have psychological aid just after his adoption to cope with some emotional problems. His 1999 MRI did not showed any sign of brain damage (Fig. 10.11).

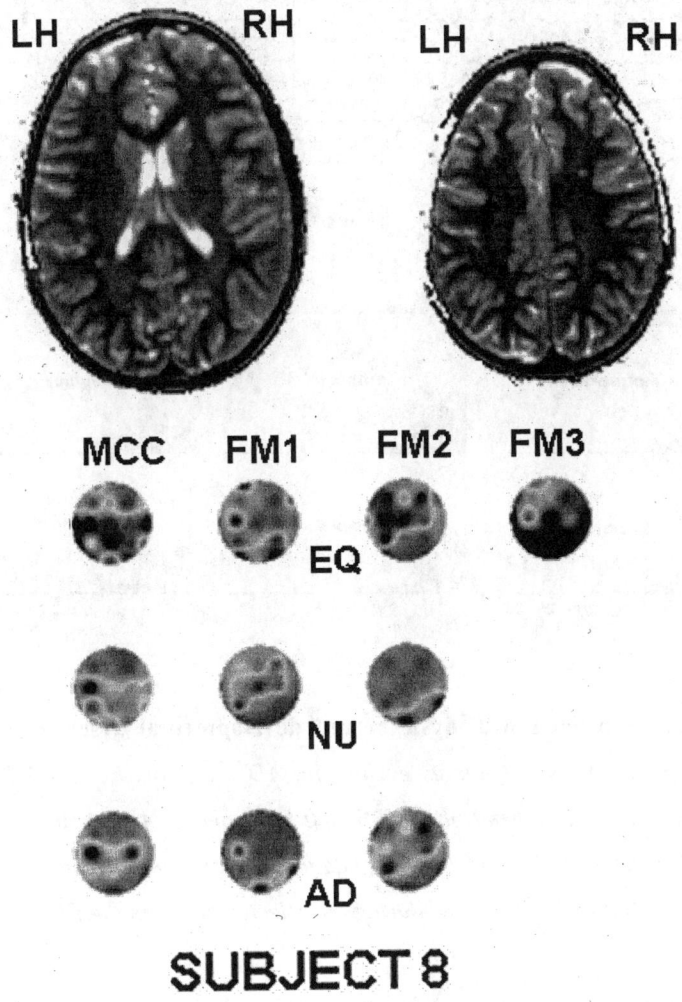

Fig. 9.10a. An abnormal MRI ...

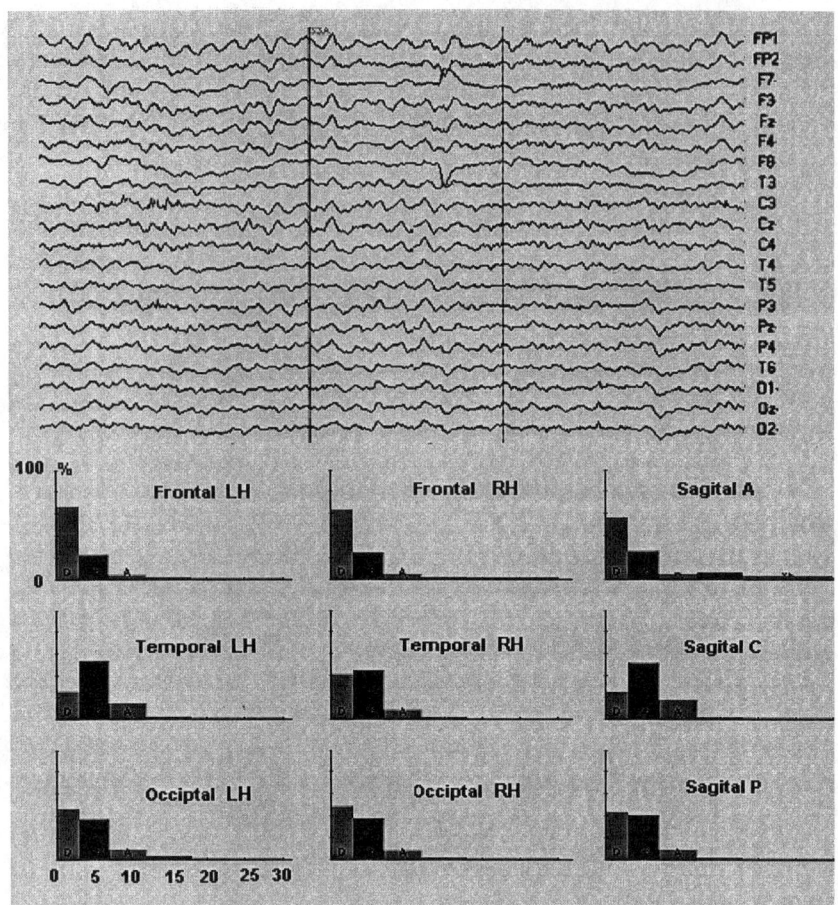

Fig. 9.10b. ...associated with dyslexia and developmental dyscalculia. *EQ: Quantity comparison; NU: Quantity evaluation; AD: Addition. For the MRI and classical EEG: RH: right hemisphere; LH: left hemisphere; A: anterior, C: central; and P: Posterior. Plot of the percentage of classic frequencies (DELTA, THETA, ALFA and BETA) obtained during the selected epoch in the EEG.*

Subject 9 started elementary school in 2001 and clearly showed signs of dyslexia and developmental dyscalculia. His main difficulty in arithmetic lay in his incapacity to deal with ordinals. He has to use a rule to sequentially order the numbers up to the desired figure. Given an arithmetic operation (e.g., addition of one digit numbers), he is able to obtain the solution, but he has to use the rule to identify the numeral associated with the quantity in the solution. His EEG was also recorded twice in 2000 and revealed in both instances the presence of Intermitent Rhythmic Delta Activity (IRDA). During one of these recording session, when solving one addition problem (9 + 4) he explicitly made a verbal request for help in visually identifying the number 13, saying: Where is the number thirteen? His FMs showed two patterns of cerebral activity association in the task of quantity/number association (NU in Fig. 9.11) and the usual three FMs in the case of addition (AD in Fig. 9.11).

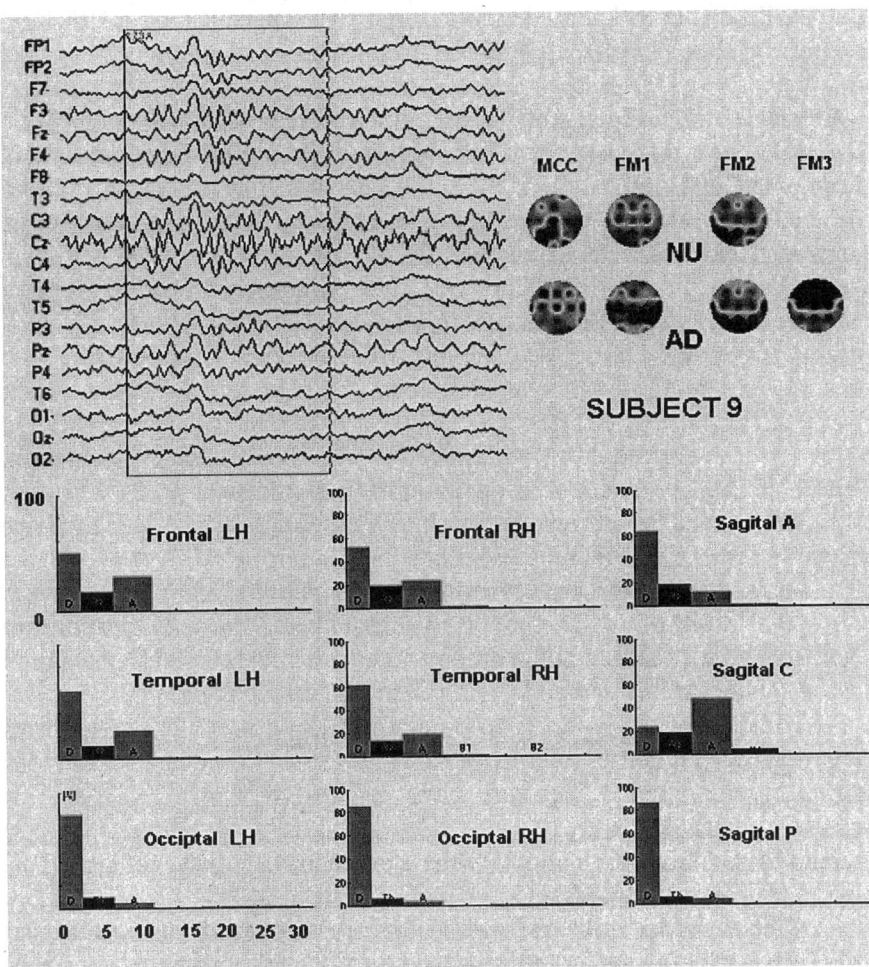

Fig. 9.11. A case of dyslexia and developmental dyscalculia with a normal MRI

Developmental dyscalculia has been defined as a pathology of otherwise normal-IQ children that is most often diagnosed when the children begin elementary school (Ansari and Karmilof-Smith, 2002; Suresh and Sebastian, 2000). Dyscalculia in low-IQ children is called *numeracy deficit* and generally believed to be genetic in origin (e.g., Ansari and Karmilof-Smith, 2002). Developmental dyscalculia is poorly understood, albeit an important issue in teaching. Only recently, Isaacs et al. (2001) have stated that, using voxel-based morphometry, they have been able to demonstrate that there is an area in the left parietal lobe where children without a deficit in calculation ability have more grey matter than those who do have this deficit. To our knowledge, this is the first report establishing a structural neural correlate of calculation ability in a group of neurologically normal individuals.

Dyslexia and developmental dyscalculia are the main deficits of subjects 8 and 9. The analysis of their EEG revealed the presence of Intermittent Rhythmic Delta Activity (IRDA). In the case of subject 8, MRI revealed a delayed myelinization, whereas in the case of subject 9, it was classified as normal. Abnormal theta activity and IRDA have been associated with dyslexia (Harmony et al. 1996; Klimesch, 1999; Klimesch et al. 2001; Rippon and Brusnwick, 2000). IRDA can be seen in association with a wide variety of pathological processes, varying from systemic toxic to metabolic disturbances to focal intracranial lesions (Cornelis and Pritchard, 1999; Neufeld et al. 1999; Sharbrough, 1999). IRDA is likely to be associated with the development of widespread brain dysfunction (Sharbrough, 1999). IRDA in subject 8 is predominantly frontal and associated with signs of delayed neural maturation. In the case of subject 9, IRDA is recorded over all electrodes, but associated with a normal MRI. Also, subject 9 exhibits dissociation between ordinal and cardinal numbers, whereas deficits of subject 8 involve both types of numbers. This seems to confirm once more the existence of the proposed components N_f and N_{cp} that are in charge of controlling and performing counting, respectively.

Harmony et al. (1996) have proposed that delta power increases in tasks that require attention to internal processing, or what they have called "internal concentration". They also published results showing that the increase in delta power was enhanced by task difficulty. Klimesch (1999) proposed that EEG oscillations in the alpha and theta band reflect cognitive and memory performance in particular if a double dissociation between the types of EEG response (tonic versus phasic) in two different EEG frequency ranges (in the alpha and theta range) is taken into account. He assumed that this dissociation is due to the fact that phasic band power increases in the theta, but decreases in the alpha frequency range, and that the extent of a phasic EEG response depends at least in part on the extent of tonic power, but in opposite ways for the theta and alpha frequency range. The delta tonic power increases under conditions that are associated with reduced cognitive processing capacity, and this augmentation of the tonic power decreases the phasic response. He also proposed that a phasic increase in a narrow frequency band of 2 Hz widths, lying below the individually defined alpha frequency, reflects encoding processes of a working memory system (Klimesch et al. 2001). Recently, Klimesch et al. (2001) showed that in the low theta band, the phasic power recorded at frontal sites did not differ when dyslexic and normal children were compared; but that registered at occipital electrodes in the dyslexic was smaller than that recorded in the normal children. These types of results clearly show that the EEG can display signals associated with abnormal cognitive processes in children, like dyslexia and dyscalculia, as described here for subjects 8 and 9. Also, dyslexia and dyscalculia may have an important genetic component (Ansari and Karmiloff-Smith, 2002; Hynd et al., 1998; Lindsay, 1996). Perhaps in some types of developmental dyscalulia a disregulation of gene reading may impair the construction of CBN circuits even while preserving much of the KFN circuits.

9.5 Conclusion

A final point to make here concerns the widespread belief in our culture in the dependence of arithmetic capability on language skills. For instance, the *triple-code model* introduced by Dehaene and colleagues (Cochon et al. 1999; Dehane and Cohen, 1995; Dehaene et al. 1999; Spelke and Tsvikin, 2001), proposes that crisp arithmetic knowledge is verbally formatted, whereas approximate calculation is quantity dependent. Although most of the children studied here experienced important language acquisition delays, no clear correlation is observed between their actual arithmetic skills and language achievements. For instance, subject 2 never exhibited any important language impairment, but her acquisition of arithmetic skills was delayed, although at the moment they have attained a high level. Subject 1 began his first phrases at age seven, but he started to master addition before he acquired a minimal writing and reading capability. Currently, his arithmetic capability is much more advanced than his oral, visual, or motor language skills. Subject 6 uttered her first worlds by the age of five, but developed very good oral language before she started to master arithmetic. Even now, her reading and writing skills are much more developed than those for calculations. Subject 8 and 9 did not experience any important impairment of their oral language development, but they are dyslexic and exhibit very defined arithmetic deficits.

The differences between our modeling of the arithmetic brain and that proposed by Dehaene et al. (Cochon et al. 1999; Dehane and Cohen, 1995; Dehaene et al. 1999; Spelke and Tsvikin, 2001) may be understood because we are currently studying the process of arithmetic knowledge acquisition by the school-aged children, whereas they based their conclusion on studies of adults, who have lost their arithmetic capabilities due to brain lesions, or on groups of bilingual adults. We believe that the innate character of the arithmetic brain strongly supports our view that arithmetic and verbal capability are founded on independent neural circuits, although language is the most important interface for humans to share information about quantities and the manipulations they perform on these quantities. Even Chomsky (Hauser et al., 2002) has recently assumed that human language capability must be a consequence of a more general brain evolution.

10 Learning Arithmetic: Why So Difficult?

The journey taken in the previous chapters on brain function let us begin to understand how the brain is genetically organized to do both fuzzy (KFN) and crisp (CBN) arithmetic, and how culture influenced humankind in creating a sophisticated theory of numbers.

10.1 The Nature of Arithmetic Knowledge

Nature selected animals with an increasing capacity for quantifying their environment in order to better survive in a hostile world, where they have to get food and avoid predators in order to maintain their bodies and to find sexual partners for the procreation of new offspring. Man inherited this knowledge (see Chap. 1) by inheriting the genetic instructions required to build up specific (CBN and KFN) neural circuits composed of specialized neurons (see Chap. 7). These mechanisms of genetic inheritance were understood and formalized in Chaps 2 and 3, where neurons were described as formal entities specialized in processing subsets of a class of fuzzy formal languages, supported by a Self-Controlled Grammar, which in turn is a class of the R Grammars (Chap. 2). In this context, the brain is proposed to be a Distributed Intelligent Processing System for such types of grammars, and to take profit from evolutionary strategies in learning how to use knowledge encoded by these languages in order to increase the odds of surviving in a continuously changing world (Chap. 4). The computational capacity of the brain greatly increased when nature discovered quantum computing and implemented it in some special cell components (Chaps. 5 and 6). In the same way that genes allow information to flow from individuals to individuals to continually reproduce and renew *genetic models* defining the different species, memes are proposed that allow information to flow from individual to individual(s) of a community so as to continually reproduce and renew culturally encoded models (Chap. 6). Memes were demonstrated to be supported by specific neural circuits processing a defined set of sentences from a fuzzy grammar (Chaps. 3 and 6). In this way, in the many instances in which mankind discovered how to increase the capacity of genetically encoded neural quantification circuits, memes were used to spread this knowledge among their fellows, who shared the same social interests (see Chaps. 6 and 7). The process of promoting such meme diffusion evolved from simple imitation to instruction (Fig.10.1), when brain processing achieved a degree of complexity allowing animals living together in a community, and to guide their fellows in repeating the same actions they have discovered by themselves, to better solve their needs (Chap. 7). The invention of language by mankind allowed instruction to evolve into teaching (Fig 10.1).

Fig. 10.1. The evolution of meme transmission: *the same meme...different ways of transmission*

Learning arithmetic is nowadays a process of developing inherited CBN circuits under the guidance of teaching. It develops as an ordered process that must begin with the construction of many different circuits in distinct cerebral areas (Chaps. 7 and 8), triggered by the questions posed in a set of problems of increasing complexity. First, it is necessary to recognize, to compare, and to order quantities and dimensions, and to identify order in sequences of events. Also, it is necessary to visually recognize the numerals. Whenever these capabilities are attained, then it is time to learn to associate numerals with cardinalities, orderings, and, finally, dimensions. To write and/or to name (but not necessarily both) the numerals is a necessary condition in communicating the results of any quantification or serial identification. The basics of addition and subtraction may be learned while learning to recognize cardinalities, if the results of such operations may be signalized with sets of corresponding cardinalities. Later, after the association of numerals with cardinalities is learned, the capability to add and to subtract can be enhanced by taking profit from formal algorithms supported by the knowledge of number as a formal structured system, and by the understanding of the formal notation used to implement such algorithms. After all these steps, the understanding of multiplication and division as special cases of counting by multiples, or as special cases of repeated additions and subtractions, can be easily attained. Such complex, ordered learning can be achieved at early ages, and as soon as it begins, gender differences are clearly established (Chap. 8) confirming the hypothesis of its innate foundations and the distributed character of numerical processing in the brain (Chaps. 8 and 9). This kind of learning may be delayed but not impeded by cerebral lesions experienced early in the life (Chap. 9).

However, if the basic neural circuits for both fuzzy and crisp numbers are genetically inherited by man, allowing us to invent many number codes on different occasions and in spatially and temporally different cultures... *why is learning arithmetic considered a hard task in all cultures?*

Several variables have been blamed as sources of stress in learning arithmetic (e.g. Dal Vesco, 2002); among them are:

1. the process of evaluating children's performance and its relative value in different cultures;
2. teaching methodologies;
3. teacher-pupil relationships;
4. peer relationships; and
5. difficulties in grasping mathematical concepts as presented by the teacher.

In the present chapter, we discuss the possible discrepancies between such factors and the brain physiology discussed in previous chapters as a source of disturbance and stress in the processes of formal arithmetical knowledge acquisition.

10.2 The Invention of Crisp Numbers

The invention of numbers by man was achieved whenever the complexity of human relations increased and pushed the development of CBN circuits to support the required arithmetic transactions.

It may be assumed that the human newborn, like any other animal, is initially equipped with KFN circuits because natural selection made the expression of their genes predominate over those required to define the CBN circuits. Perhaps this is because the complexity of the arithmetic transactions for survival is easily (and maybe better) solved by fuzzy arithmetic, supported by KFN circuits. Therefore, the human invention of numbers will demand, first of all, the enhancement of the expression of the genes encoding CBN agents, and then the development of adequate connections among them. DIPS learning of a new task may require both the creation of new agents and the establishment of defined commitments among them.

Let k_{max} and c_{max} be, respectively, the maximum number of KFN and CBN circuits that a given brain may build. Let also k_a and c_a be, respectively, the number of KFN and CBN circuits that a given brain has actually built. Finally, let

$$k = k_a/k_{max} \text{ and } c = c_a/c_{max} \qquad (10.1)$$

Now, let α be defined in the closed interval [0,1] to measure the relative gene expression for building KFN and CBN agents, such that

$$\text{if } \alpha \rightarrow 1 \text{ then the KFN gene expression prevails} \qquad (10.2)$$

otherwise

$$\text{if } \alpha \rightarrow 0 \text{ then the CBN gene expression prevails} \qquad (10.3)$$

In such conditions:

Conjecture 10.1. The newborn capability $\rho(CBN \mid G,H)$ for inventing crisp arithmetic may be calculated as:

$$\rho(CBN \mid G,H) = 1 - (\alpha * k)/c, \ 0 < \alpha < 1 \qquad (10.4)$$

Conjecture 10.2. It is assumed here that in the case of animals other than man:

$$k \gg c \text{ and } \alpha \rightarrow 1 \qquad (10.5)$$

whereas in the case of the human newborn

$$k > c \text{ and } \alpha \to 0 \tag{10.6}$$

In this condition, in the case of animals

$$\rho(CBN \mid G,H) < 0 \tag{10.7}$$

whereas in the case of humans

$$\rho(CBN \mid G,H) > 0 \tag{10.8}$$

Remark 10.2. Conjectures 10.1 and 10.2 are proposed to explain the large difference in arithmetic capability between human and animals, despite the fact that the grammar G defining their brains shares the genes required by both KFN and CBN circuits. On the one hand, as α and k/c increases, the chances an animal creates a CBN circuit decreases and ultimately becomes impossible. On the other hand, the difficulty a human may have inventing (or learning) crisp arithmetic is mainly determined by the actual value of the relation k/c. People suffering specific brain lesions may have this value increase, and thus having $\rho(CBN \mid G,H)$ decrease, because of the loss of CBN agents. But since they did not lose the genes defining these agents, they retain the capability for recreating such circuits, as has been shown in Chap. 9. Genetic disturbances may either reduce α or increase k/c. But even in these conditions, learning will be possible if CBN gene expression is not totally abolished.

Conjectures 10.1 and 10.2 also stress the influence of the environment H upon the human capability for creating different crisp number systems and for doing so on different occasions, as well as in cultures that are spatially and temporally isolated.

10.3 Learning by Observing

A series of very interesting experiments was performed by Petrosini et al. (2003) to study the properties of rats learning by observing other rats in solving a given problem. They investigated how rats may learn to find a submerged platform in a swimming pool (Fig. 10.2).

When a rat is put in a swimming pool (Fig. 10.2 left), it tries to escape:

1. swimming around the borders to jump out of the pool. After a few attempts, it discovers that this strategy is not a solution for its problem, and then

2. randomly swims across the pool and incidentally discovers the platform, where it may find a safe place. After that, it improves its performance by
3. quickly learning how to find the platform.

Now, if another rat is first allowed to observe the first rat discovering how to find the safe place in the swimming pool (Fig. 10.2 right), and then put in the pool, it quickly finds the platform, demonstrating that it has learned to solve the problem by observing the behavior of the first rat. This clearly shows the meme diffusion in rodents, promoted by imitation.

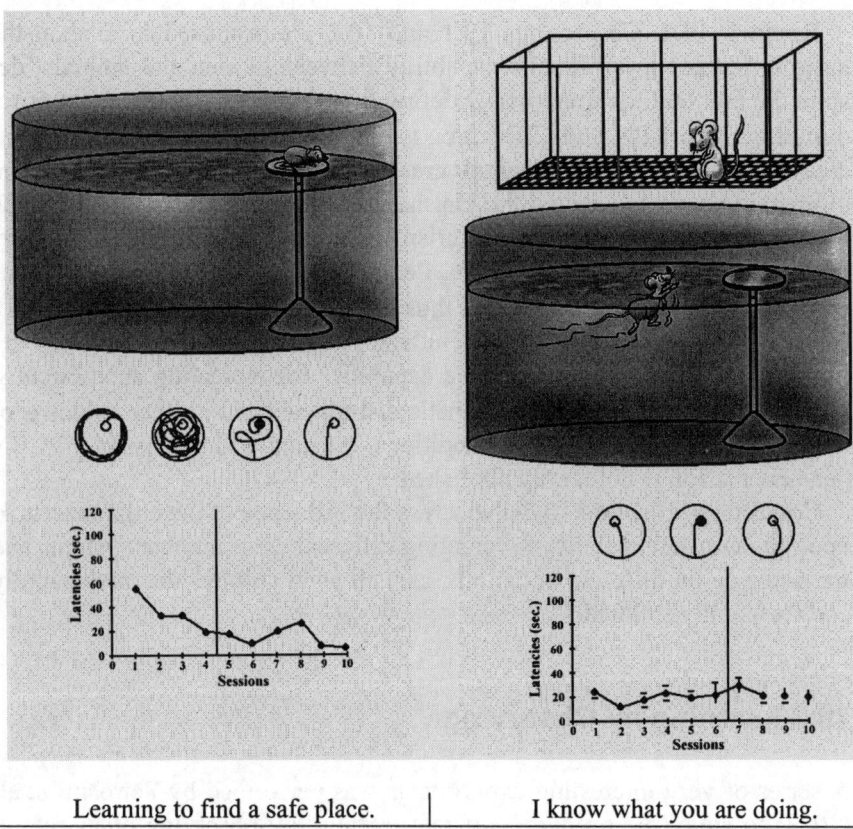

Learning to find a safe place. | I know what you are doing.

Fig. 10.2. Learning by observation: *I know what you are doing!*

The interesting results from these experiments arose when the second rat was allowed to observe only one of the phases of the learning being experienced by the first rat (Fig. 10.3). In all cases, the partial observation resulted in blocking the normal learning of the second rat. It not only failed to profit from observing the first rat, as in the normal condition, (Fig. 10.3 upper row), but the incomplete observation of the task solution resulted (Fig 10.3 middle and low rows) in very poor learning!

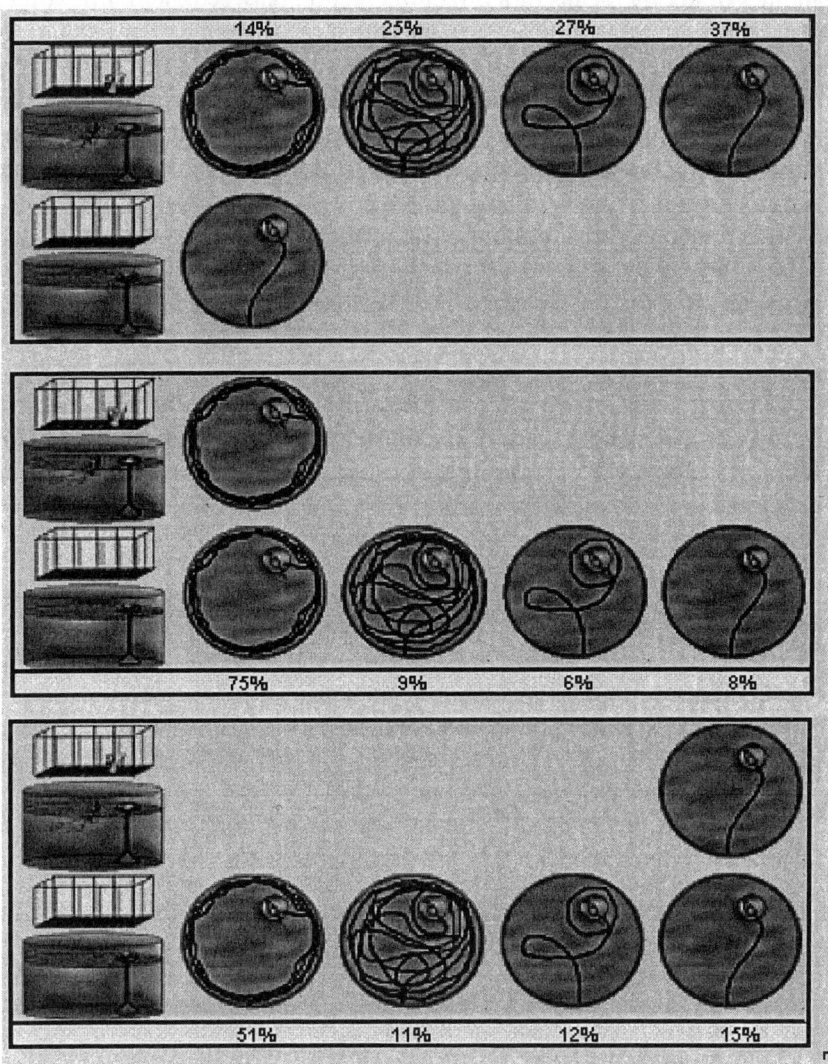

Fig. 10.3. If you do not correctly show me how...I can't learn! *Poor teaching...*

10.4 Arithmetic Meme Diffusion in School

As discussed and modeled in Chap. 6, meme diffusion is dependent on two main communication modalities.

In the first, mailing system, a meme active in one brain is replicated in another brain, by means of a direct interaction between them. Imitation and instruction are the two processes by which meme replication occurs. Here, imitation implies intentional action of the receptor, but not necessarily of the transmitter, whilst instruction implies intentional action of both actors. A good example of instruction is the tutorial teaching system, the basis of guild training in the Middle Ages that is still in use in some academic areas such as medicine and graduate courses.

In the second, broadcasting system, memes are hardware-stored in the culture shared by a group of actors, and are intentionally searched for by these actors. Probably the first broadcasting hardware is the painted cave wall, like *Pedra Furada* in Brazil (Chaps. 1 and 2), and the latest being the Internet.

Teaching relies on both mechanisms, because it has the purpose of replicating knowledge stored in the culture using the teacher as a mediator of such replication. The meme in the culture is first revived in the teacher's brain in order to begin replication in the pupil's brain. The teacher is responsible for selecting the memes to be replicated, and then helping the student to handle them.

As discussed in Chaps. 7 through 9, there exist in the brain at least three systems of number, supported by different neural circuits:

1. the ordinals: supported by ranking operations over serial events, performed by some specific frontal circuits;
2. the integers: supported by quantification of set cardinality, performed by defined frontal and parietal circuits, and
3. the reals (R): supported by quantification operations over the three continuous space dimensions: length, width and high.

Here must be added a fourth system:

4. currency numbers: supported by adding emotional valuation to integer and/or real quantities of trading interest.

As discussed in Chap. 7, abstract concepts may be constructed by discovering common properties shared by similar well-developed neural circuits. In this line of reasoning, higher numerical concepts are constructed from the learning of common properties of the number systems listed above. In this way, man created/discovered the notion of an abstract number system and number theory by learning about the common properties of at least four (a through d, above) different number neural circuits. Such a theory is comprised of a collection of related memes and it is this theory that is available in present human culture. In this context, crisp number representations and properties (number theory) and their operations (group theories), evolved from the more primitive knowledge represented in the neural circuits depicted above. Note that the construction of the numerical knowledge base in human culture occurred via an evolutionary process similar to the evolutionary knowledge construction where the rat learns to find a safe place in the water maze. The difference between the two processes is in the time scale. Whereas culture involves knowledge evolution from generation to generation, individual knowledge evolution occurs during the individual life span.

In this line of reasoning, two learning strategies may be considered in the acquisition of crisp number representation and their operations by school-age children. On one hand, the evolutionary process of arithmetic knowledge acquisition by children must recapitulate the initial steps of the evolutionary pathway of cultural development of number and group theories under the teacher's guidance. This is a process similar to that of the rat learning by observing how to do it. On the other hand, children must recreate the main steps of universal arithmetic evolution under the influence of an adequate environment. This is a process similar to that of the rat learning by doing it.

It is proposed here that learning arithmetic is so difficult because children are not shown how to develop their innate circuits and how to evolve the abstract arithmetic embodied in our culture from this innate knowledge. A clear example (Bransford et al., 2003, pp. 92) of how teachers make arithmetic learning difficult is illustrated in Fig.10.4, involving the confusion they create between ordinal and cardinal numbers. One of the most popular number exercises is to copy, write, or name the numbers in sequences from 1 to ..., as a good exercise in learning the quantities represented by them. The teacher's rationale is that the children will be associating the numerals (1, 2, 3, ...) to the quantities to be represented by them, because each time the child copies, writes, or names the numeral, he/she will be adding one to the initial quantity in the series. However, this type of exercise is interpreted by the child as one of identifying the elements of a series, which is to say, learning the ordinal numbers. In this kind of exercise, the child uses each element numeral as a label of one of the elements of the series, as for example, the first, the second ..., etc. To mistake two different innate neural circuits not only imposes a burden on the children in the initial learning of arithmetic, but it will also contribute to making any future learning more difficult, as in the case of not correctly showing the rat how to solve the water-maze.

Fig. 10.4. If you do not correctly show me how to do it ... I can't learn! *Making wrong assumptions about the child's reasoning.*

Subjects 6, 7 and 9 described in Chap. 9 are examples of initial poor teaching. Despite their cerebral lesions the children were able to develop their ordinal and cardinal numbering circuits. But they were over-trained to use the ordinals to communicate arithmetical operations with the cardinals. Since their lesions rendered the ordinal circuit less efficient than the cardinal circuit, they had great difficulty telling the teacher about the results of their calculations. One of them (subject 6) invented a process of her own to solve the problem (see Fig. 9.5). She wrote down the numerals serially under each element of the sets involved in the calculation in order to discover how to tell others the result computed by her cardinal neural circuit. Subject 9 asked many times for some help in finding the numerals in a series in order to relate his calculations. Once, he asked one of the authors how to write the number *thirteen,* which was the correct result of the addition problem (9 + 4) he had solved.

10.5 Evolving Arithmetic Knowledge in the School

It may be assumed that man started to develop our actual crisp numbers and arithmetic gradually, as the complexity of human society began to increase.

The first attempts, on the cave walls (see Fig. 6.2 and Fig. 10.5), and with bones, were one-to-one mapping between the objects to be counted and the elements of representation. The next steps involved shortening such notations, for instance in the case of the Sumerian clay jar, recording commercial transactions, which quantities were encoded by one-to-one mapping between a pebble inside the jar and a corresponding symbol written outside. This was the beginning of an abstract number notation. Arithmetic calculation began to be computed by handling pebbles in a primitive abacus (e.g., by the Greeks and Romans) or rope knots (e.g., the quipua of the Incas), etc. Only very late in the history of mathematics, with the introduction of Arabic numerals and the Indian zero, did modern algorithms begin to be used to solve arithmetic calculations (Ifrah, 1985, Joseph, 1990).

Pre-school children have a good knowledge of number and calculation, as shown in their mimicking the first one-to-one mapping representations and calculations (Fig. 10.5). By the time children begin school, most have built a considerable knowledge store relevant to arithmetic. They have experiences of adding and subtracting numbers of items in their everyday play, although they lack the symbolic representations for addition and subtraction that are taught in school (Fig. 10.5a) (Bransford et al., 2003). Teaching our formal number notations and arithmetic calculations with small numbers may be facilitated by mimicking the Sumerian double procedure, that is, pairing one-to-one mapping procedures with formal number notation (Fig. 10.5b). Also, calculations may be easily learned if primitive (Greek or Roman) abaci are allowed (Fig. 10.5c). By following this path, children start to understand symbolic notation and become ready to learn the algorithmic approach to arithmetic calculation (Fig. 10.5d). This evolutionary pedagogy simulates the history of number knowledge acquisition, such that neural circuits for abstract numbers begin to be naturally created in the brain as the common properties of the different initial CBN circuits, which are then developed and correlated with the symbolic notation.

Under the current pedagogy, instead of taking this evolutionary approach to arithmetic learning, the young child is frequently asked to compare set cardinalities or evaluate dimensions using linguistic variables before he/she is allowed to work with crisp quantification. For instance, the child has to select the set having few, many ... elements, or he/she has to say who is tall, short, etc. Children are asked to strengthen their KFNs in order to be able to create their CBNs. But from conjectures 10.1 and 10.2, this reduces $\rho(CBN \mid G, H)$, and makes arithmetic learning much more difficult.

Fig. 10.5 a,b,c,d: Evolution of arithmetic knowledge: *Ontogeny repeats phylogeny?*

10.5 Evolving Arithmetic Knowledge in the School

The contemporary view of learning is that people construct new knowledge and understanding based on what they already know and believe. However, if students' initial ideas and beliefs are ignored, the understanding that they develop can be very different from what the teacher intends. Thus, teachers must draw and work with the preexisting knowledge that their students bring with them (e.g., Bransford et al., 2003) and not try to model the child's brain according to their adult knowledge.

It is our proposition, here, that if teachers take into account the physiology of the arithmetic neural circuits, as described and discussed in this book, in planning their classroom activities, they surely will make arithmetic learning a more pleasant and easy journey.

References

Ahn J, Weinahct TC, Bucksbaum PH (2000) Information storage and retrieval through quantum phase. *Science* 287: 463-465

Alkon DL, Nelson TJ, Zhao W, Cavallaro S (1998) Time domains of neuronal Ca2+ signaling and associative memory: steps through a calexcitin, ryanodine receptor, K+ channel cascade. *Trends Neurosci* 21 (12), 529-537

Anderson RM, May RM (1991) Infectious Diseases of Humans. Oxford Science Publications, Oxford

Anholt RR (1994) Signal integration in the nervous system: adenylate cyclases as molecular coincidence detectors. *Trends Neurosci* 17: 37-41

Ansari D, Karmiloff-Smith A (2002) Atypical trajectories of number development: a neuroconstructivist perspective. *Trends Cogn Sci* 12: 511-516

Akutsu, TS Kuhara O Maruyama M and Miyano S (2003) Identification of genetic networks by strategic gene disruptions and gene overexpressions under a Boolean model. *Theoretical Computer Sciences* 298: 235-251

Arshavsky YL (2002) Cellular and network properties in the funcitoning of the nervous system: from central pattern generators to cognition. *Brain Research Review*, 41 (2-3): 229-267

Ashcraft MH (1991) Cognitive arithmetic: a review of data and theory. In: Dehaene, S (ed) Numerical Cognition, Blackwell, Oxford

Babloyantz A (1986) Molecules, Dynamics and Life: an Introduction to Self-Organization of Matter. John Wiley and Sons, New York

Barbieri M (2003). *The Organic Codes: an introduction to semantic biology.* Cambridge University Press. Cambridge

Barth HN, Kanwisher N, Spelke E (2003) The construction of large number representations in adults. *Cognition*, 86:201-221

Beal C R, (1999) Special Issue on the Math-Fact Retrieval Hypothesis. *Comtemp. Educ. Psychol.* 24 (3): 171-180

Bennett CH, Divincenzo DP (2000) Quantum information and computation. *Nature* 404, 247-255

Bennett CH, Shor PW (1999) Privacy in the quantum world. *Science* 284: 747

Bickle J, Avison M, Schmithorst V (2003) Bridging the cognitive-cellular neuroscience gap empirically: a study combining physiology, modelling and fMRI *J Exp Theor Artif In* 15 (2): 161-175

Blackmore SJ (1997) Probability misjudgement and belief in paranormal: a newspaper survey. *British Journal of Psychology* 88: 683-689

Blackmore S (1999) *The Meme Machine.* Oxford University Press, Oxford

Bliss TVP, Collingridge GL (1993) A synaptic model of memory: long-term potentiation in the hippocampus. *Nature* 361: 31-39

Bouwmeester D, Zeilinger A (2000) The physics of quantum information: basic concepts. In: Bouwmeester D, Ekert A, Zeilinger A (eds) The Physics of Quantum Information. Springer-Verlag, Berlin

Boyd R, Richerson P J (1985) Culture and the evolutionary process. University of Chicago, Chicago

Brannon, EM (2002) The development of ordinal numerical knowledge in infancy. *Cognition* 83 (3): 223-240

Brannon EM, Terace HS (1998) Ordering of the numeoristies 1 to 9 by monkeys. *Science* 282:746

Bransford J, Brown AL, Cocking RR (2003) How People Learn: Brain, Mind, Experience and School. National Research Council, Washington

Brassard G, (1997) Searching a quantum phone book. Science 275: 627-628

Brassard G, Chuang S, Lloyd S, Monroe C. (1998) Quantum computing. *Proc Natl Acad Sci* USA 95: 11032-11033

Bray NJ, Buckland PR, Owen MJ, et al. (2003) Cis-acting variation in the expression of a high proportion of genes in human brain. *Hum Genet* 113 (2): 149-153

Brunet M (2002) Palaeoanthropology: Sahelanthropus or 'Sahelpithecus'? Reply. *Nature* 419 (6907): 582-582

Butterworth, B (1999) The mathematical brain, Macmillan Publishers, London

Cabeza R, Nyberg L (1997) Imaging cognition: an empirical review of PET studies with normal subjects. *Journal of Cognitive Neuroscience*, 9: 1-2

Cabeza R, Nyberg L (2000) Imaging cognition II: an empirical review of 275 PET and fMRI studies. *Journal of Cognitive Neuroscience*, 12: 1-47.

Carpenter, AF, Georgopoulos AP, Pellizzer G (1999) Motor cortical encoding of serial order in a context-recall task. *Science*, 283: 1752-1757

Cavalli-Sforza L, Feldman W (1981) Cultural transmission and evolution: a quantitative approach. Princeton University Press, Princeton

Chandrasekaran B (1981) Natural and social system metaphors for distributed problem solving: introduction to the issue. *IEEE Transation System Man and Cybernetics*, 11: 11-15

Changeux JP, Dehaene S (2000) Hierarchical neuronal modeling of cognitive functions: from synaptic transmission to the Tower of London. *Intern Jnl Psychophysiol* 35 (2-3): 179-187

Chomsky, N (1957). *Syntactic Structures*. The Hague, Mouton

Chomsky, N (1995). *Aspects of the Theory of Syntax*. MIT Press, Cambridge

Chuang II, Gershenfeld N, Kubinec M (1998) Experimental implementation of fast quantm searching. *Physical Ver. Letters* 80/15: 3408-3411

Chugani HT (1999) Metabolic imaging: A window on brain development and plasticity. *Neuroscientist* 5: 29-40

Cirac JI, Zoller P (2000) A scalable quantum computer with ions in an array of microtraps. *Nature* 404: 579-581

Cochon E, Cohen L, van de Moortele W, Dehaene S (1999) Differential contributions of the left and right ingerior parietal lobules to number processing. *J Cognitive Neurosci* 11: 617-630

Conroy GC (1997) Reconstructing Human Origins: a modern synthesis. WW Norton & Company, N. York

Cornelis JS, Pritchard WS (1999) Dynamics underlying rhythmic and non-rhythmic variants of abnormal, waking delta activity. *Int J Psychophysiology* 24: 5-20

Cowel SF, Egan GF, Code C, Harasty J, Watson JDG (2000) The functional neuroanatomy of simple calculation and nmber repetition: A parametic PET activation study. *NeuroImage*, 12: 565-573

Dal Vesco AA (2000) Alfabetizacão Matemática e as Fontes de Estresse no Estudante. Editora da Universidade de Passo Fundo, Passo Fundo

Davis H, Perusse R (1988) Numerical Competence in Animals: definitional issues, current evidence and a new research agenda. *Behav Brain Sci* 11 (4): 561-579

Davis R, Smith RG (1983) Negotiation as a metaphor for distributed problem solving. *Artificial Intelligence*, 20: 63-109.

Dawkins R (1976) The Selfish Gene. Oxford University Press, Oxford

Dehaene S (1991) Varieties of numerical abilities. In: Dehaene S (ed) Numerical Cognition. Blackwell, Oxford

Dehaene S (1997) The Number Sense: How the Mind Creates Mathematics. Penguin Books, London

Dehaene S (2002) Single-Neuron arithmetic. *Science* 297: 1652-1653

Dehaene S (2003) The neural basis of the Weber-Fechner law: a logarithmic mental number *Trends Cogn Sci* 7 (4): 145-147

Dehaene S, Dehanene-Lambertz G, Cohen L (1998) Abstract representations of numbers in the animal and human brain. *Trends Neuroscience* 21: 355-361

Dehaene E, Spelke E, Pinel P, Stanescu R, Tsvkin S (1999) Sources of mathematical thinking: behavioral and brain-imaging evidence. *Science*, 284: 970-974

Devlin K (2001) The Maths Gene. Phoenix, London.

Edelman GM (1989) The Remembered Present: a biological theory of consciousness. Basic Books, New York

Edelman GM (1997) Neural Darwinism: the theory of neuronal group selection. Basic Books, New York

Edmonds B (1998) On modelling in memetics. *Journal of Memetics* http://jom-emit.cfpm.org/1998/vol2/edmonds_b.html

Enard W, Przeworski M, Fisher SE (2002) Molecular evolution of FOXP2, a gene involved in speech and language. *Nature* 418: 869-872

Fayol M (1996) A Criança e o Número. Artes Médicas, Porto Alegre

Feigenson L, Carey S, Hauser M (2002) The representations underlying infants' choice of more: object files versus analog magnitudes. *Psychol Sci* 13 (2): 150-156.

Feldman, MW Laland, KN (1996) Gene-culture coevolutionary theory. *TREE*. 11: 453-457

Ferber J (1999) Multi-Agent Systems: An Introduction to Distributed Artificial Intelligence. Addison-Wesley, Harlow

Fink GR, Marshall JC, Gurd J, Weiss PH, Zafiris O, Shah JJ, Zilles K (2001) Deriving numerosity and shape fron identical visual displays. *Neuroimage* 13: 46-55

Fox MS (1981) Reorganizational view of distributed systems. *IEEE Transaction System Man and Cybernetics* 11: 70-80

Fox K (2003) Synaptic Plasticity: The Subcellular Location of CaMKII Controls Plasticity, Current Biology, Volume 13: 4/R143-R145

Foz FB, Lucchini FP, Palimieri S, Rocha AF, Rodella EC, Rondó AG, Cardoso MB, Ramazzini PB, Leite CC (2002) Language Plasticity Revealed by EEG Mapping. *Pediatric Neurology*, 26: 106-115

Gabora L (1997) The Origin and Evolution of Culture and Creativity. *Journal of Memetics - Evolutionary Models of Information Transmission:* http://jom-it.cfpm.org/vol1/gabora_l.html

Gallager AM, De Lisi R, Holst PC, McGillicuddy-De Lisi AV, Morely M, Cahalan C (2000) Gender differences in advanced mathematical problem solving. *J Experimental Child Psychology* 75: 165-190

Gallistel CR (1990) Representations in Animal Cognition: an introduction. *Cognition* 37 (1-2): 1-22

Gallistel CR, Gelman R (1991) Preverbal and verbal counting and computation. In: Dehaene, S (ed) Numerical Cognition. Blackwell, Oxford

Gallistel CR, Gelman R (2000) Non-verbal numerical cognition: from reals to integers. *Trends in Cog. Sciences*, 4: 59-65

Gatherer D (2001) Modelling the effects of memetic taboos on genetic homosexuality. Journal of Memetics http://jomemit.cfpm.org/2001/vol4/gatherer_d.html

Gelman R, Meck E (1983) Preschooler's couting: principles before skills. *Cognition*, 13: 343-359

Gertzmann J (1924) Figeragnosie und isolierte Agraphie; ein neues Syndrom. *Ges. Neurol. Psychiat.* 108:152-177

Ghosh A, Greenberg ME (1995) Calcium signaling in neurons: molecular mechanism and cellular consequences. *Science* 268: 239-247

Gilmore G (1995) Alice in Quantumland. Springer-Verlag, Heidelberg

Gobel S, Walsh V, Rushworth MFS (2001) The mental number line and the human angular gyrus. *NeuroImage*, 14: 1278-1289

Groen JG, Parkman JM (1972) A chronometric analhysis of simple addition. *Psychological Rev* 79: 329-343

Guidon N (1998) As ocupações pré-históricas do Brasil. In: Manuela Carneiro da Cunha (org) História dos Índios do Brasil. Companhia das Letras, São Paulo

Halpern DF (1992) Sex differences in cognitive abilities, 2nd edn. Lawrence Erlbaum Associates, Hillsdale

Harmony TT, Fernández J, Silva J, Bernal L, Díaz-Comas A, Reyes E, Marosi M, Rodríguez M (1996) EEG delta activity: na indicator of attention to internal processing during performance of mental tasks. *Int J Psychophysiology* 24: 161-171

Harris RK (1996) *Encyclopedia of Nuclear Magnetic Resonance.* Vol. 5. John Wiley & Sons, Chichester/UK

Hauser M D, Chomsky N, Tecumseh W, Fitch C (2002) The Faculty of Language: What Is It, Who Has It, and How Did It Evolve? *Science* 298: 1569-1579

Helmchen F (2002) Raising the seepd limit – fast Ca^{2+} handling in dendritic spines. *Trends in NeuroSciences* 25, 438-441

Hewitt C, Inman J (1991) DAI betwixt and between: From "Intelligent Agents" to Open System Science. IEEE Transation System Man and Cybernetics, 21:1409-1419

Hinrinchs JV, Yurko DS, Hu JM (1981) Two digit number comparison: use of place information. *J. Exptl. Psychology: Human Perception and Performance* 7:890-891

Hodgkin AA, Huxley AF (1952) A quantitative description of membrane currents and its application to conduction and excitation in nerve. *J Phsyiology*, 116:500-544

Holtoff K, Tsay D, Yuste R (2002) Calcium dynamics of spines depend on their dendritic location. *Neuron* 33: 425-37

Hopcroft JE, Ullman JD (1969) Formal Language and the relation to automa, Addison-Wesley, Reading, MA

Huntley-Fenner G (2001) Children's understanding of number is similar to adults 'and rats': numerical estimation by 5-7 year olds. *Cognition*, 78 b27-b40

Hynd G W, Hooper SR, Takahashi T (1998) Dyslexia and Language-Based disabilities. In: Coffey and Brumbak (eds) Text Book of Pediatric Neuropsychiatry. American Psychiatric Press, pp. 691-718

Ifrah G (1985) Les chiffres ou l'historie dúne grande invention. Éditions Robbert Laffont AS, Paris.

Isaacs EB, C. J. Edmonds CJ, Lucas A, Gadian DG (2001) Calculation difficulties in children of very low birthweight: a neural correlate. *Brain*, 124/9: 1701-1707

Igushi Y, Hashimoto I (2000) Sequential information processing during mental arithmetic is reflected in the time course of event-related brain potentials. *Clinical Neurohysiology*, 111: 204-213

Jahanshahi M, Dirnberger G, Fuller R, Frith CD (2000) The role of the dorsolateral prefrontal cortx in randmom nuber generation: A study with Positron Emission Tomography. *Neuroimage*, 12: 713-725

Jensen EM, Reese EP, Reese TW (1950) The subtizing and couting of visually presented fields of dots. *The Journal of Psychology* 30: 363-392

Jonides J (1995) Working Memory and Thinking. In: Smith EE, Osherson DN (eds) An Invitation to Cognitive Science, Vol. 3: Thinking. MIT Press, Cambridge

Jonides J, Smith EE (1997) The Architecture of Working Memory. In: Rugg MD (ed) Cognitive Neuroscience. MIT Press, Cambridge

Joseph GG (1990) The Crest of the Peacock: Non-European Roots of Mathematics. Penguin Books, London

Journal of Memetics Evolutionary Models of Information Transmission 4: http://jomemit.cfpm.org/2000/vol4/kendal_jr&laland_kn.html

Harrison WA (2001) Applied Quantum Mechanics. World Scientific, Singapore

Kasai H (2003) Structure-stability-function relationships of dendritic spines. *Trends in Neurosciences*

Kaufman EL, Lord MN, Reese TW, Volkmann J (1949) The discrimination of visual number. *American J of Psychology* 62: 498-525

Kawai N and T. Matsuzawa (2000) Numerical memory span in a chimpanzee. *Nature* 403 (6765): 39-40

Kendal JR, Laland KN (2000) Mathematical Models for Memetics. *Journal of Memetics*. http://jom-emit.cfpm.org/2000/vol4/kendal_jr&laland_kn.html

Kielpinski D, Monroe C, Wineland DJ (2002) Architecture for a Large-Scale Ion-Trap Quantum Computer. *Nature* 417: 709-711

Klimesch W (1999) EEG alpha and theta oscillations reflect cognitive and memory performance: a review and analysis. *Brain Res Rev* 29:169-195

Klimesch W, Doppelmayr M, Wimmer H, Schwaiger J, Röhm D, Bruber W, Hutzler F (2001) Theta band power changes in normal and dyslexic children. *Clinical Neurophysiology* 113:1174-1185

Knight RT (1997) Distributed cortical network for visual attention. *Journal of Cognitive Neuroscience* 9/75-91

Kong J, Wang Y, Shang H, Wang Y, Yang X, Zhuang D. (1999) Brain potentials during mental arithmetic — effects of problem difficulty on event-related brain potentials. *Neurosciences Letters*, 260: 169-172

Konig P, Engel AK, Singer S (1996) Integrator or Coincidence Detector? The role of the cortical neuron revisited. *Trends in Neurosciences* 19(4): 129-136

Kostich M, English J, Madison V, Gheyas F, Wang L, Qiu P, Greene J, Laz TM (2002) Human members of the eukaryotic protein kinase family. *Genome Biology* 3(9): 0043.1-0043.12
http://genomebiology.com/2002/3/9/research/0043

Krystal JH, Belger A, D'Souza C, Anand A, Charney DS, Aghajanian GK, Moghaddam B (1999) Therapeutic implications of the hyperglutamatergic effects of NMDA antagonists. *Neuropsychopharm* 21 (56) 143-57

LaBerge D (2001) Attention, consciousness, and electrical wave activity within the cortical column. *Int J Psychophysiol* 43 (1): 5-24

LeDoux , J. (2002) The Synaptic Self. Penguin Books, New York pp 174-199

Lesser VR (1991) A retrospective view of FA/C distributed problem solving. *IEEE Transation System Man and Cybernetics* 21:1347-1362.

Lieshoff C, Bischo HJ (2003) The dynamics of spine density changes. *Behavioural Brain Research*, 140, 87-95

Lindsay RL (1996) Dyscalcuia. In: Capute and Accardo (eds) Developmental Disabilities in Infancy and Childhood. Paul Brookes Publishing, Baltimore, pp 405-415

Lisman J, Schulman H, Cline H (2002) The molecular basis of CaMKII function in synaptic and behavioural memory. *Nat Rev Neurosci* 3(3):175-90

Lynch, A. (1998) Units, events and dynamics in memetic evolution. *J. of Memetics*: http://jom-emit.cfpm.org/1998/vol12/lynch_a.html

Marks J (2002) What it means to be 98% chimpanzee. University of California Press, Berkeley

Massad E, Rocha AF (2002) Meme-Gene coevolution and cognitive mathematics. In: Abe JM, Silva Filho JI (eds) Advances in Logic, Artificial Intelligence and Robotics. IOS Press, Amsterdam, pp 75-81

Maunsell JHR, Ferrera VP (1995) Attentional mechanisms in visual cortex. In: Gazzaniga MS (ed) The Cognitive Neurosciences. The MIT Press, Cambridge, pp 451-461

Mayer E, Martory MD, Pegna AJ, Landis T, Delavelle J (1999) A pure case of Gertzmann syndrome with a subangular lesion. *Brain*, 122: 1107-1120

McCloskey M (1991) Cognitive mechanisms in numerical processing: Evidence from acquired dyscalculia. In: Dehaene S (ed) Numerical Cognition, Blackwell, Oxford

McCloskey MW, Harly SM, Sokol, P. (1991) Models of arithmetic fact retrieval. An evaluation in light of findings from normal and brain-damaged subjects. J Exptl Psycholog: *Learning, Memory and Cognition*, 17: 377-397

McCulloch WS, Pitts W (1943) A logical calculus of the ideas immanent in nervous activity. *Bull. Mathematical Biosphysics* 5: 115-133

Mechner F (1958) Sequential Dependencies of the Lengths of Consecutive Response Runs *J Exp Anal Behav* 1 (2): 229-233 1958

Meck WH, Church RM (1983) A mode control model of couting and timing processes. *J Exptl. Pshychol Anim Behav Proc* 9: 320-334

Menon VS, Rivera M., White CD, Eliez S, Glover GH, Reiss AL (2000) Functional optimization of arithmetic processing in perfect performers. *Cog Brain Research* 9: 343-345

Miller KD (2003) Understanding Layer 4 of the cortical circuit: A model based on cat V1. *Cerebral Cortex*, 13,73-82

Mix KS (1999) Preschoolers's recognition of numerical equivalence: Sequential sets. *J Exptl Child Psychology* 74: 309-332

Mizumoto, M., Toioda, Y and Tanaka, K. (1973) N-Fold Grammar. *Information Sciences*, 5, 22, 1973

Monod J, Changeux JP, Jacob F (1963) Allosteric proteins and cellular control systems. *J. Mol. Biol.* 6: 306-329

Nagerl HC, Leibfried D, Schmidt-Kaler F, Eschner J, Blatt R, Brune M, Raimond, JM, Haroche S (2000) Cavity QED-Experiments: Atoms in Cavities and Trapped Ions. In: Bouwmeester D, Ekert A, Zeilinger A (eds) The Physics of Quantum Information. Springer-Verlag, Berlin

Navarro EA (1998) Método Moderno de Tupi Antigo. Editora Vozes, Petrópolis

Negoita, C.V. and Ralescu, D.A. (1975). *Applications of Fuzzy Sets to Systems Analysis*. John Wiley & Sons, New York.

Neufeld MY, Chistik V, Chapman J, Korczyn AD (1999) Intermittent rhythmic activity morphology cannot distinguish between focal and diffuse brain disturbances. *J Neurological Sciences* 164:56-59

Nieder A, Freedman DJ, Miller EK (2002) Representation of the quantity of visual items in the primate prefrontal cortex. *Science* 297: 1708-1711

Nieder A, Miller EK (2003) Coding of cognitive magnitude: compressed scaling of numerical informaton in the primate prefrontal cortex. *Neuron*, 37: 149-157

Nielsen MA, Chuang IL (2000) Quantum Computation and Quantum Information. Cambridge University Press, Cambridge

Nuerk H-C, Weger U, Willmes K (2001) Decade breaks in the mental number line? Putting the tens and units back in different bins. *Cognition*, 82: B25-B33

Pedrycz W, Lam PCF, Rocha AF (1995) Distributed Fuzzy System Modeling. *IEE Trans. Sys. Man Cyber* 25: 769-780

Pedrycz W, Gomide F (1998) An Introduction to Fuzzy Sets. Bradford Books, Cambridge

Pereira Jr. A (2002) Neuronal Plasticity: how memes control genes. In: Abe JM, Silva Filho JI (eds) Advances in Logic, Artificial Intelligence and Robotics, IOS Press, Amsterdam, pp 82-87

Pereira Jr. A (2003) The Quantum Mind/Classical Brain Problem. *Neuroquantology* 1: 94-118: http://med.ege.edu.tr/~tarlaci/2003_Vol%2001/content.htm

Pereira Jr. A, Johnson G (2003) Towards an Understanding of the Genesis of Ketamine-Induced Perceptual Distortions and Hallucinatory States. *Brain and Mind* 4 (3), in press.

Pereira Jr A, Rocha AF (2000) Temporal Aspects of Neuronal Binding, In: Buccheri R, Soniga M, e Gesu V (eds) Studies in the Structure of Time: from Physics to Psychopathology, Kluwer, New York

Pesenti M, Thioux M, Seron X, De Volder A (2000) Neuroanatomical substrates of Arabic number processing, numerical comparison and simple addition: A PET study. *J Cognitive Neurscience* 12: 461-479

Petrosini L, Graziano A, Mandolesi L, Neri P, Molinari M, Leggio MG (2003) Watch how to do it! New advances in learning by observation. *Brain Res Ver* 42: 252-264

Piazza M, Mecheli A, Butterworth B, Price CJ (2002) Are subtizing and counting implemented as separate for functionally overlapping processes? *Neuorimage*, 15: 435-466

Platt JR, Johnson DM (1971) Localization of position within a homogeneous behavior chain: Effects of error contingencies. *Learning and Motivation* 2: 386-414

Polk T A, Reed CL, Keenan JM, Hogarth P, Anderson CA (2001) A dissociation between Symbolic Number knowledge and Analogue Magnitude Information. *Brain and Cognition*, 47: 545-563

Posner M (1995) Attention in Cognitive Neuroscience. In: Gazzaniga MS (ed) The Cognitive Neurosciences, MIT Press, pp 615-624

Ratinck R, Brysbaert M, Reynvoet B (2001) Bilateral field interactions and hemispheric asymmetry in number comparison. *Neuropyshologia*, 39: 335-345

Rickard TC, Romero SG, Basso G, Wharton C, Flitman S, Grafman J (2000) The calculating brain: na fMRI study. *Neuropsychologia*, 38: 325-335

Rippon G, Brunswick N (2000) Trait and state EEG indices of information processing in developmental dyslexia. *Int J of Psychophysiology* 36: 251-265

Rizzolatti G, Arbibi M (1998) Language within our grasp. *Trends in Neurosci* 21: 188-194

Rocha AF, Françozo E, Hadler MI, Balduino MA (1980) Neural languages. *Fuzzy Sets and Systems* 3:11-35
Rocha AF (1982a) Basic properties of neural circuits. *Fuzzy Sets and Systems*, 7: 109-121
Rocha AF (1982b) Toward a theoretical and experimental approach of fuzzy learning. In: Gupta MM, Sanches E (eds) Fuzzy Reasoning and Approximate Decision, North Holland, Dordrecht pp191-202
Rocha AF (1992) Neural Nets: A theory for brains and machine. *Lecture Notes in Artificial Intelligence,* Springer-Verlag, Berlin
Rocha AF (1997) The brain as a symbol-processing machine. *Progress in Neurobiology* 53, 121-198
Rocha AF, Rebelo MPF, Miura K (1998) Toward a theory of molecular computing. *Information Sciences* 106: 123-157
Rocha AF, Pereira Jr. A, Coutinho FAB (2001) NMDA channel and consciousness: from signal coincidence detection to quantum computing. *Progress in Neurobiology* 64, 555-573
Rocha AF, Massad E (2002) Evolving Arithmetical Knowledge in a Distributed Intelligent Processing System. In: Abe JM, da Silva Filho JI (ed) Advances in Logic, Artificial Intelligence and Robotics Frontiers in Artificial Intelligence and Applications 85: 68-74.
Rocha AF, Massad E (2003a) How the human brain is endowed for mathematical reasoning: Are numbers the product of evolution of brain on Earth or have they an existence of their own? *Mathematics Today*, June 2003: 15 - 18
Rocha AF, Massad E (2003b) Ambiguous Grammars, Genetic Variability and Evolution (Submitted)
Rocha AF, Leite CC, Rocha FT, Massad E, Cerri GG, Angelotti SAO, Gomes EHG (2003a) Mental retardation: A MRI study of 146 Brazilian children (Submitted)
Rocha AF, Rocha FT, Massad E, Leite C (2003b) Brain plasticity and arithmetic learning in congenitally injured brains. (Submitted)
Rocha FT, Rocha AF, Massad E (2003c) Brain mappings of the arithmetic processing in children and adults. (Submitted)
Rocha FT, Rocha AF, Massad E (2003d) Arithmetic reasoning: experiments and modelling. (Submitted)
Rogers E (1962) Diffusion of Innovation, Free Press, New York
Rogers E (1995) Diffusion of Innovation, 4th edition. Free Press, New York
Royer JM, Tronsky L (1999) Math-Fact retrieval and gender. *Contemporary Educational Psychology* 24: 181-266
Sabatini BL, Maravall M, Svoboda K (2001) Ca2+ signaling in dendritic spines. *Current Opinion in Neurobiology*, 11/3:349-356
Sabatini BL Thomas G. Oertner and Karel Svoboda (2002) The Life Cycle of Ca2+ Ions in Dendritic Spines, *Neuron,* 33/3: 31: 439-452
Sacks O (1985), The man who mistook his wife for a hat. Duckworth, London
Sadegh-Zadeh K (2000) Fuzzy genomes. *Artificial Intelligence in Medicine*, 18: 1-28

Sawamura H, Shima K, Tanji (2002) Numerical representation for action in the parietal cortex of the monkey. *Nature*, 415: 918

Schmidt-Kaler F, Haffner H, Riebe M, Guide S, Lancaster GPT, Deuschie T, Becher C, Roos CF, Eschner J, Biatt R (2003) Realization of the Cirac-Zoller Controlled-NOT Quantum Gate. *Nature*, 422: 408-411.

Searls DB (2002) The Language of Genes. *Nature*, 420:211-217

Searls, DB (1993) The Computational Linguistics of Biological Sequences. In: Hunter L (ed) Artificial Intelligence and Molecular Biology. AAAI Press / MIT Press, Cambridge pp 46-120

Shalev RS, Gross-Tsur V (2001) Developmental dyscalculia. *Pediatric Neurology*, 24: 337-342

Sharbrough FW (1999) Nonspecific Abnormal EEG Patterns. In: Niedermeyer K, Lopes da Silva F (eds) Electroencephalography: Basic Principles, Clinical Applications and Related Fields. Williams & Wilkins, Baltimore pp 215-234

Shettleworth SJ (1998) Cognition, Evolution and Behavior. Oxford University Press, Oxford

Shuresh PA, Sebastian S (2000) Developmental Gerstmann's syndrome: A distinct clinical entity of learning disabilities. *Pediatric Neurology*, 22: 267-278

Siegler RS (1996) Emerging minds. Oxford University Press, Oxford

Skrandies W, Reik P, Kunze C (1999) Topography of evoked brain activity during mental arithmetic and language tasks: sex differences. *Neuropsychologia*, 37: 421-430

Smythies J (2002) Brain story: Unlocking our inner world of emotions, memories, ideas and desires. *Psychol Med* 32 (2): 374-375

Spelke ES, Tsvikin S (2001) Language and number: a bilingual training study. *Cognition*, 78: 45-88

Stahian K, Simon TJ, Peterson S, Patel GA, Hoffman JM, Grafton ST (1999) Neural evidence linking visual object enumeration and attention. *J Cognitive Neuroscience* 11: 36-51

Stanescu-Cosson R, Pinel P, van de Moortele PF, LeBihan D, Cohen L, Dehaene S (2000) Understanding dissociations in dyscalculia. L A brain imaging study of the impact of number size on the cerebral networks for exact and approximate calculation. *Brain* 123: 2240-2255

Starkey P Cooper RG (1980) Perception of Numbers by Human Infants. *Science* 210 (4473): 1033-1035

Strauss MS Curtis LE (1981) Infant perception of numerosity. *Child Development* 52: 1146-1152

Sweatt JD (1999) Toward a molecular explanation for long-term potentiation. *Learn Mem* 6: 399-416

Tattersall I, Schwatz J (2000). Extinct Humans. Westview Press.

Thompson PM, Gield JN, Woods RP, MacDonald D, Evans AC, Toga AW (2000) Growth patterns in the developing brain detected by using continuum mechanical tensor maps. *Nature* 404: 190-193

Tuszynski, J. A. and M. Kurzynski (2003) Introduction to Molecular Biophysics. CRC Press, Boca Raton, pp.229-275

Verkhratskt A (2002) The encodolasmic reticulum and neuronal calcium signaling. *Cell Calcium* 32, 393-404

Vianna MR, Alonso M, Viola H, Quevedo J, de Paris F, Furman M, de Stein ML, Medina JH, Izquierdo I (2000) Role of hippocampal signaling pathways in long-term memory formation of a nonassociative learning task in the rat. *Learn Mem* 7 (5): 333-40

Villablanca JR, Hovda DA (2000) Developmental neuroplasticity in a model of cerebral hemispherectomy and stroke. *Neuroscience* 95: 625-637

Warren SW (1997) The usefulness of NMR Quantum Computing. *Science* 277: 1668-1669.

White TD; Asfaw B, DeGusta D, Hilbert H, Richards GD, Suwa G, Howell FC (1997) Pleistocene Homo sapiens from Middle Awash Ethiopia. *Nature* 243: 742-747

Williams, RJP, Fráusto da Silva, JJR (2003) Evolution was chemically constrained. *J. Theor. Biol.* 220:323-343

Wolpoff MH, Senut B, Pickford M, et al. (2002) Palaeoanthropology - Sahelanthropus or 'Sahelpithecus'? *Nature* 419 (6907): 581-582

Wynn K (1996) Infants' individuation and enumeration of actions. *Psychol Sci* 7 (3): 164-169

Wynn K (1998) Psychological foundations of number: numerical competence in human infants. *Trends in Cognitive Sciences* 1998: 122-131

Wynn K (2000) Early numerical knowledge, In: Lee K (ed) Childhood cognitive development, Blackwell, Oxford

Woodruff G, Premack D (1981) Primitive Mathematical Concepts in the Chimpanzee: proportionality and numerosity. *Nature* 293 (5833): 568-570

Xu F, Spelke ES (2000) Large Number Discrimination in 6-Month-Old Infants *Cognition* 74 (1): B1-B11

Zadeh LA (1965) Fuzzy sets. *Information and control*, 8: 338-353

Zago LM Pesenti E, Mellet F, Crivello B, Tzourio-Mazoyer N (2001) *NeuroImage* 13: 314-327

Zeilinger A (1998) Quantum Entanglement: A Fundamental Concept Finding Its Applications. *Physica Scripta* T76: 203-209

Zorzi M, Priftis K, Umilta C (2002) Neglect disrupts the mental number line. *Nature* 417: 138-139

Zupan, B., I. Bratko, J. Demsar, P. Juvan, T. Curk, U. Brostnik, J. Robert Beck, J. Halter, A. Kuspa and G. Shaulsky (2003) GenePath: a system for inference of genetic networks and proposal of genetic experiments. *Artificial Ingelligence in Medicine* 29:107-13

Index

acceptance, 28,29,63,72,74
accumulator, 139, 141,144,146
adapted, 43, 44, 74
addition, 146, 171
adenylyl cyclase, 83,84
agnosia, 16
allosteric mechanisms, 82
ambiguity, 26,30,35
AMPA, 84,89,91
amplitude interference, 113
amplitudes, 80,113
attainability, 71,72
attention control, 62
attention deficit, 18
axonic transportation system, 49

Bell states, 80
bipedalism, 18,19
blackboard posting, 65
broadcasting, 122
Brocca's area, 16
buffer, 115
bursts, 225

Ca^{2+} pumps, 90
calmodulin, 84,91,96
calmodulin-dependent protein kinase, 86,91
cardinalities, 31,201
cardinality quantification, 138
cellular specialization, 22
central nervous system, 60
central-parietal (N_{cp}) cells, 172
chimpanzees, 9,12,18
code string, 22
coincidence-detectors, 82,84,110

commitment, 156,162
communicating, 62
competition, 76,79
continuous spiking, 100
control string, 23
controlled NOT, 81
cooperation, 79
copying-fidelity, 118
counting, 54,117,118,131
crisp base numbers, 142
currency numbers, 206

decision, 71
decoherence, 112, 225
degree of acceptance, 29,63
degree of similarity, 29
dendritic spine, 89,91,94
Deutsch-Josza algorithm, 93
developmental dyscalculia, 18,196
developmental Gerstmann's syndrome, 179
DIPS reasoning, 54,61
division, 148
dyscalculia, 22,192
dissociation, 16,189
distributed intelligent processing system, 4,21,53
distributed intelligent processor, 53
division, 148,168,174
double dissociation, 216, 196
dyscalculia, 18
dyslexia, 18,194,196

EEG mapping, 153
electric code, 49
electrical gradient, 47
electro-encephalography, 17

embriogenesis, 57,77
endoplasmatic reticulum, 90,95
enrollment capability, 156
Enscer figure database, 105
entanglement, 79,80,87
entropy, 33,83
enumerate, 7,15
evaluating, 62,72
evolution, 19,33,43
evolutionary learning, 70,75
executive processor, 115,116
expressiveness, 33,35,51

fecundity, 118
frontal-parietal gliosis, 189
fuzzy alphabet, 29
fuzzy base number system, 142
fuzzy formal language theory, 30
fuzzy grammar G, 31,35
fuzzy numbers, 136,192

game event related activity, 159
gene mutation, 38
genetic network, 25,39
Gertzmann syndrome, 16,191
glutamate, 24 ,96
glycyne, 83
goal definition, 62

Hadamard gate, 81
Hadamard transformation, 98,99
hominids, 18,123
hormones, 22,65
hyperactivity, 18

implementing, 62
imprinting, 93,104,106
inducer substrings, 24
inferior parietal, 9,13,153
innate numerical competence, 9
intelligence, 54
intermitent rhythmic delta activity, 192
internal states, 87,91
interneurons, 100
ion-trap computers, 87,91

K fuzzy numbers, 136
kinase, 84,85

laser pulses, 88,94
lateral prefrontal, 12,114
layers, 100
learning by observing, 61,203
learning capability, 67,179
left frontal neurons, 171
leukomalacy, 180
local response, 48,49
longevity, 118
looking time, 14

magnesium, 83
Magnetic Resonance Imaging, 17
Magneto-Encephalography, 17
mail addressing, 64,122
Man-Whitney U test, 164
matching, 28
mean ambiguity, 33,107
meme-gene-coevolution, 118
memes, 117
memetics, 117,126
memorizing, 62
memory span, 12
mental calculation, 136,154
mental lexicon, 17
mental number line, 139
microtraps, 87
model, 70
molecular neurobiology, 21
motor cortex, 12,191
multi-organ, 42,47,57
multiplication, 148

natural selection, 13,15
nervous system, 59
neural network models, 55
neuronal specialization, 54
neurons (N_v), 172
neurotransmitters, 65
new language, 44,68
nicotinic acetylcholine receptor, 84
NMDA receptor, 83,96

nuclear magnetic resonance, 87,93
number, 13,15,20
number grammar, 17
number sense, 137,150
numbers, 117
numeracy deficit, 179,196
numerosity, 11,13,149

ordinal neurons, 141
organ, 42,56
oscillator, 88

parallel fibers, 100
parietal number representation, 189
Pedra Furada, 124,206
phasic, 100,196
phonons, 88
planning, 62,71
plasticity, 17,76,129,191
Positron Emission Tomography, 17
possibility, 31,32,52
post-synaptic density, 95,96
prefrontal cortex, 12,114,123,153
principal components analysis, 161
proteome, 22
pyramidal cells, 100,102

quadripole, 88,91,94
quantifiers, 142,149
quantum bit, 80
quantum charge-coupled computers, 93
quantum computation, 3,79,93
quantum computer, 81,88
quantum cortical pattern recognition device, 93
quantum cryptography, 87
quantum gates, 81
quantum information, 87,93
quantum superdense coding, 113

radio frequency pulses, 87
recurrent processing, 109
recursive enumerable grammars, 36
replication, 118,206
repressor substrings, 24

retrograde signal, 50
rewriting, 28,138

schyzencephaly, 188
second-order catalysts, 83,109
selective attention, 110,115
self-controlled grammar, 36,55,121,199
Sensor, 105
sequential ordering, 153
signal transduction pathways, 24
somato-sensory cortex, 13
specialized agents, 54,79
spike, 49
spine pruning, 104
stellate, 100,102
subtizing, 9,.133,138
subtraction, 148,164
superfamilies, 61
superposition, 79,80,87,112
switch, 91
synchronous oscillations, 110

TATA box, 23
tonic, 100,197
transgenic mice, 89
travelling salesman problem, 140
triple-code model, 135,197
two-level systems, 87
two-photon microscopy, 89
type 0 grammar, 28

unitarity, 92

ventral frontal cortex, 115
vibrational states, 87
visual pattern recognition device, 100
voltage-dependent calcium channels, 93

waveform, 87
Weber's law, 8
working memory, 114

zero, 15

Printing: Strauss GmbH, Mörlenbach
Binding: Schäffer, Grünstadt

DUE DATE SUBJECT TO CHANGE
IF A RECALL IS REQUESTED

DEC 2 8 2007

Rtnd-AB JAN 1 4 2008